The Internet of Things

How Smart TVs, Smart Cars, Smart Homes, and Smart Cities Are Changing the World

MICHAEL MILLER

800 East 96th Street,
Indianapolis, Indiana 46240 USA

The Internet of Things

How Smart TVs, Smart Cars, Smart Homes, and Smart Cities Are Changing the World

ISBN-13: 978-0-7897-5400-4
ISBN-10: 0-7897-5400-2

Library of Congress Control Number: 2015932632

Printed in the United States of America

First Printing: April 2015

Trademarks

Warning and Disclaimer

Special Sales

For information about buying this title in bulk quantities, or for special sales opportunities (which may include electronic versions; custom cover designs; and content particular to your business, training goals, marketing focus, or branding interests), please contact our corporate sales department at corpsales@pearsoned.com or (800) 382-3419.

For government sales inquiries, please contact governmentsales@pearsoned.com.

For questions about sales outside the U.S., please contact international@pearsoned.com.

Cover image credit: thermostats–shanesabin/Shutterstock | city–Sebastian Kaulitzki/Shutterstock | earth–Zffoto/Shutterstock | car–AlexRoz/Shutterstock | house–ekaterina7/Fotolia | quadcopter–frog/Fotolia | smart tv–viperagp/Fotolia | fridge–You can more/Fotolia | drone–mipan/Fotolia | abstract square pattern–marigold_88/Fotolia

Editor-in-Chief
Greg Wiegand

Executive Editor
Rick Kughen

Managing Editor
Sandra Schroeder

Senior Project Editor
Tonya Simpson

Copy Editor
Anne Goebel

Indexer
Erika Millen

Proofreader
Jess DeGabriele

Technical Editor
Gareth Branwyn

Publishing Coordinator
Kristin Watterson

Cover Designer
Alan Clements

Compositor
Nonie Ratcliff

CONTENTS AT A GLANCE

TABLE OF CONTENTS

3   Smart TVs: Viewing in a Connected World 43

4   Smart Appliances: From Remote Control Ovens to Talking Refrigerators 61

7   Smart Shopping: They Know What You Want Before You Know You Want It 145

About the Author

Michael Miller has written more than 150 nonfiction how-to books over the past two decades, as well as a variety of web articles. His best-selling books include Que's *Absolute Beginner's Guide to Computer Basics, The Ultimate Guide to Bitcoin,* and *Is It Safe? Protecting Your Computer, Your Business, and Yourself Online.* Collectively, his books have sold more than 1 million copies worldwide.

Miller has established a reputation for clearly explaining technical topics to non-technical readers and for offering useful real-world advice about complicated topics. More information can be found at the author's website, located at www.millerwriter.com. His Twitter handle is @molehillgroup.

Dedication

To my six wonderful grandchildren, who will inherit the future we're creating— Collin, Alethia, Hayley, Judah, Lael, and Jackson.

Acknowledgments

Thanks to the usual suspects at Que, including but not limited to Rick Kughen, Greg Wiegand, Tonya Simpson, Anne Goebel, and technical editor Gareth Branwyn.

We Want to Hear from You!

As the reader of this book, *you* are our most important critic and commentator. We value your opinion and want to know what we're doing right, what we could do better, what areas you'd like to see us publish in, and any other words of wisdom you're willing to pass our way.

We welcome your comments. You can email or write to let us know what you did or didn't like about this book—as well as what we can do to make our books better.

Please note that we cannot help you with technical problems related to the topic of this book.

When you write, please be sure to include this book's title and author as well as your name and email address. We will carefully review your comments and share them with the author and editors who worked on the book.

Email: feedback@quepublishing.com

Mail: Que Publishing
           ATTN: Reader Feedback
           800 East 96th Street
           Indianapolis, IN 46240 USA

Reader Services

Visit our website and register this book at quepublishing.com/register for convenient access to any updates, downloads, or errata that might be available for this book.

- Black & white images are crap
- gray background contrast is dreadful

Introduction

You've probably heard about the Internet of Things, sometimes called the Internet of Everything. You might not know what it is (and, frankly, the definition is a little fuzzy), but you've heard about it and you're interested in it enough to pick up this book. Good for you.

Like you, I was curious about the Internet of Things (which we'll abbreviate to IoT from here on out). I wasn't quite sure about what it was or where I could find it or even what it consisted of. All I knew is that everybody was talking about it, in the tech world at least, and thus it attracted my attention.

So, as is my wont, I went out and learned about the IoT. Then I wrote about what I learned, and the result is the book you hold in your hands, The Internet of Things: How Smart TVs, Smart Cars, Smart Homes, and Smart Cities Are Changing the World. Read along and you'll learn as I did what this IoT thing is all about.

Spoiler alert: It isn't quite as clear cut as you'd think by the name. Yes, the Internet of Things is literally about things connected to the Internet, but it's both more and less than that.

In many ways, the IoT is marketing hype, a buzz phrase used to describe all manner of new devices and services that various manufacturers would very much like for you to purchase. There are a lot of companies adding the word "smart" to the devices they sell in the hope of tagging along on the IoT bandwagon. That's to be expected; remember all the "cyber" and "e-" things back in the early days of the Internet? Everybody wants to be on top of the latest trend. That's where the money is.

The technical definition of the IoT involves small devices, each with their own Internet Protocol (IP) address, connected to other such devices via the Internet. In other words, lots of little "things" connected to lots of other little "things" over the Internet. Instead of connecting people to other people, as does the current Internet, the new Internet of Things connects things to things. That sounds simple.

Except, a lot of the so-called smart devices ballyhooed as part of the IoT don't have their own IP addresses, don't connect to the existing Internet, and don't even connect to other devices. Which means the IoT isn't just about connecting things to things; it's also about autonomous operation—things that can operate pretty much on their own, without a lot of human interaction.

And even those devices that do connect to other devices don't connect to all other devices. A lot of what I found about the IoT involves industry-specific applications, where concepts of thing-to-thing connectivity and autonomous operation are applied to solve very specific problems. There's a distinct IoT for the healthcare industry, and another for the automotive industry, and another for the warehousing/distribution chain, and so on. The smart medical devices you'll find in your local hospital have nothing at all to do with the smart cars you might find parked in the hospital parking lot, or the smart systems employed to put food in the hospital cafeteria. Chances are they don't even use the same network to connect.

For that reason, you have to look at the IoT as multiple networks of things, each dedicated to specific industries or applications. That's how I approached it in this book, which is why you'll find separate chapters for smart homes, smart clothing, smart cars, smart medicine, and such. In a way, each of these areas will have its own Internet of Things, to which its own devices and services will be connected.

Like I said, it's not just one thing. It's lots of things.

This will all make more sense to you as you read through the book. We start with a general introduction to the IoT and its underlying technologies, then move into examinations of the many different approaches to the IoT, from the most personal (smart homes and smart clothing) to the more universal (smart medicine, smart cities, smart warfare). We end with a chapter describing the potential problems associated with the IoT, of which there are several.

By the end of the book, you should be a lot better versed in the various things that are likely to comprise the Internet of Things. And you'll know how all of this is likely to affect you, personally. It's really quite thrilling.

What You Need to Know to Use This Book

How much prior knowledge of the Internet of Things do you need before starting this book? Absolutely none. I assume that you, like me when I first got started, don't know much of anything about the Internet of Things. This isn't a really technical book, so you don't have to come into it with a bunch of detailed technical knowledge either. In other words, this book is written for—and can be read by—anyone who's curious about the IoT. If I do my job right, this book will assuage that curiosity.

One More Thing

There's one more thing you need to know about the Internet of Things before you start reading. That is this—like all emerging technologies, the Internet of Things is in the process of defining itself. There's a lot of change happening, and it's happening every single day. What I write about the IoT today may be superseded tomorrow. It's an exciting time full of rapid development and constant discoveries, so don't expect things to stay the same for long. Read this book to get a general overview of what's happening, but then keep your ear to the ground to stay on top of ongoing developments.

Where can you learn more about the IoT, and find the latest news and developments? Alltop hosts a nice feed of IoT-related news, located at internet-of-things. alltop.com. So does TechCrunch, at www.techcrunch.com/topic/subject/internet-of-things/. And, for more business-oriented stories, check out Venture Beat's IoT feed at http://venturebeat.com/tag/internet-of-things/.

I'm guessing, however, that you'll find plenty of IoT-related stories in your day-to-day news reading. Like I said, it's a big buzzword, which means it's getting an increasing amount of coverage, even in the mainstream press. Just keep your eyes and ears open and you'll hear more about it.

1

Smart Connectivity: Welcome to the Internet of Things

The Internet of Things is coming. It's going to be big. It's going to be important. It's going to seriously impact your life.

But what is the Internet of Things? And why is it so big, important, and impactful? Read on to find out.

Welcome to the Future

In the future, the world will be different.

Imagine the home of the future. It's a smart home; it knows what you're doing and adjusts itself accordingly. It knows when you're on the way home from work and turns on the lights, turns up the heat, turns on the oven, and maybe even turns on your favorite streaming music station to welcome you when you walk in the door. It knows when the best times of day are to run the washer and clean the dishes; it knows to turn off the lights when you leave a room and lock the doors when you drive away.

Imagine the car of the future. Like your smart home, it's a smart car. It knows who's driving and adjusts to your driving and listening and heating/cooling pref-erences; you'll get sport handling, cooler temps, and the country station, while your spouse gets cruising mode, warmer temps, and the easy listening station. It also knows when something's wrong or needs maintenance and schedules its own appointments with the repair shop.

Imagine the fitness machines at your local gym programming themselves for your personal workout when you leave the locker room. Imagine a factory where every machine in every room feeds back information to help the production line run more efficiently. Imagine an entire city that manages its public lighting, utility delivery, and road repairs automatically, based on real-time conditions and needs.

Imagine this world of the future; then consider how it will work. It's all about con-necting all manner of things together, in an intelligent fashion. It's what many are calling the Internet of Things.

What Is the Internet of Things?

You're familiar with the Internet. It's that global network that connects millions of computers (and smartphones and tablets), enabling electronic communication between those computers. More practically, when connecting those computers, the Internet connects together the users of those computers. It's all about people using the global network to share information and messages.

So today's Internet is a network of machines, yes, but also a network of people. All those machines connected together serve the diverse purposes of their human users.

The Internet of Things, however, is different. Instead of creating a network that connects people, it's a network that connects things. Lots of things.

You see, the Internet of Things (experts call it the IoT, for short), connects not just computers and smartphones and tablets (which are all computing devices of a sort), but also lots of other things. That's the whole point of the thing, to connect just about every thing in the aptly-named Internet of Things. And once connected, every thing can communicate with every other thing for a variety of useful purposes.

Put more technically, the IoT is the interconnection of uniquely identifiable embedded computing devices. That means any device that can be connected—not just computers, but various types of sensors and monitors, too. The interconnection of said devices can take place over the existing Internet infrastructure, so it kind of piggybacks on today's Internet. (Although the connections don't have to take place over the Internet—they can use other network technologies, including proprietary wireless networks.)

Let's put it another way: Today's Internet is an Internet of People. Yes, it's the computers and smartphones and tablets (and, behind the scenes, the servers and networks) that are connected, but they're connected primarily for the use of people. People use their computing devices to access the Internet, to search for and read information, to send emails and instant messages, to download music and videos and porn and whatever else happens to be stored there. The devices are connected only to serve their human users.

The Internet of Things is in stark contrast to the Internet of People. Instead of people accessing data and communicating with one another, the IoT enables things to access data and communicate with one another. Yes, humans will continue to use the current Internet, but the future Internet will also be a pipeline for non-human devices—M2M (machine-to-machine) communication, if you will.

What happens when you connect just about any type of thing with just about any other type of thing? First off, you generate a ton of data. Whether the thing connected is a television set, refrigerator, heart monitor, or automobile, the sensors inside each thing will amass a huge amount of data about what that thing is doing and how its interacting with its environment. All that data can then be transmitted to other things (over the Internet, of course) and used to automate additional actions by those and other devices.

And that's the promise of the IoT. More, more automatic, and more intelligent services provided by interconnected smart devices—with a minimal amount of human interaction.

What Kinds of Things Can Be Connected to the Internet of Things?

So the Internet of Things is all about things. That makes sense. But what kinds of things are we talking about?

In essence, a "thing" in the IoT can be anything large enough to contain a wireless transmitter (employing Wi-Fi, Bluetooth, or any other wireless protocol) and unique enough to be assigned its own Internet Protocol (IP) address. This could include something as small as a paperclip or as large as a house.

The IoT can connect:

- Home electronics devices, such as so-called smart televisions and streaming media servers
- Medical devices, such as pacemakers and heart monitoring implants
- Home appliances, such as smart refrigerators, ovens, and laundry machines
- Automobiles—including self-driving cars
- Airplanes large and small—from commercial airliners to self-flying drones
- Home automation devices, such as thermostats, smoke detectors, and alarm systems
- Homes and towns and cities and nations—just about anything that can be monitored and controlled

Things on the IoT don't have to be inanimate, either. Animals, such as dogs, cats, and cows can be embedded and connected, as can human beings. Consider biochip transponders that keep track of wandering farm animals, or implants that monitor the physical location or medical conditions of human wearers.

What all these things have in common, in addition to their ability to connect to one another, is that they either contain a sensor or have the ability to perform a specific task—or both. One type of thing communicates to the other, and things get done.

The result is literally billions of things connected to each other via the IoT. A gigantic number of connected devices that dwarfs the number of computers and smartphones connected to today's Internet.

In fact, there will be so many individual things connected to the Internet of Things that some have taken to calling it the "Internet of Everything." That might be stretching things (so to speak) a little too far, but you get the point. Connect enough things together, and it will seem like virtually everything is connected.

 Note

> Out of all the alternative names, the appellation "Internet of Things" appears to be the one that has stuck. That's in spite of the fact that many of these devices aren't directly connected to the Internet, but rather

connected to each other (or larger networks) via other wireless protocols. For this reason, and because of the sensor technology embedded in many of these devices, some are calling this next generation of connectivity the "Sensor Revolution." Not quite as sexy as the Internet of Things, but maybe a bit more descriptive.

What Do All Those Connected Things Do?

Most of the things connected to the IoT are actually simple devices that are often referred to as *smart devices*. (As in smart TVs, smart refrigerators, or—believe it or not—smart diapers.) The devices themselves aren't necessarily smart in and of themselves, but become smart when joined together with other connected devices.

In fact, when you first look at it, there don't appear to be much that's new or unique about the things connected to the Internet of Things. We have connected devices today. We have devices with embedded sensors today. We have devices that perform one or more discrete tasks today. So it's not the devices or the sensors or the fact that they're linked together that makes the IoT so exciting.

Rather, it's the fact that once enough of these devices come together they create a coherent system that can act with its own type of intelligence, without the need for human interpretation and interaction. It's like all these relatively simple devices (simple when compared to a personal computer, anyway) combine to create a single, giant machine—much like individual bees swarming together in a hive intelligence.

In the IoT, every connected device becomes something greater than any given individual device by itself. The whole is greater than the sum of its parts, because everything is communicating with everything else in an intelligent, automated fashion. Any given device connects to other surrounding and relevant devices to share collected data. This creates what experts call *ambient intelligence*, which results when multiple devices act in unison to carry out everyday activities and tasks using the information and intelligence embedded into the network. It all happens in the background, automatically, serving people's needs without requiring people's help or interaction.

It's all about combining data collection with the ability to perform specific activities. Some connected devices contain sensors that register conditions in the world around them—temperature, light, motion, you name it. These devices transmit the collected data to other devices that are hardwired to perform specific actions. These action devices do what they're programmed to do, the results of which are measured by the sensor devices. It's a self-correcting cycle that becomes smarter over multiple iterations.

Here's an example, concerning your car. Today if something breaks in your car, a sensor recognizes that something's wrong and activates the Check Engine light. It's a very simple application of sensor technology and not entirely useful (you don't know exactly what's wrong, after all), but it's where we're at today.

In tomorrow's Internet of Things, you get more sensors communicating with other devices in a more intelligent fashion. Instead of a single sensor connected to the Check Engine light, your new smart car will sport a multitude of sensors, many built into individual car parts. When a sensor notices that a particular part is wearing out—something that would have lit the Check Engine light before—the sensor notifies a controller or "brain" mounted somewhere in your car. The controller notes the faulty part and, when the car is next connected to the Internet (typically by driving into your home Wi-Fi network zone), it sends a message to your friendly automobile repair shop. The computer at the shop looks up information about the faulty part, determines the necessary repairs, orders a replacement part (if one isn't already in stock), and contacts the calendar application on your smartphone to schedule a service appointment to replace the part. No difficult-to-fathom warning lights, no unexpected breakdowns, no scheduling hassles. All the necessary things communicate with all the other necessary things to get your car back in working condition as quickly as possible, with as little hassle as possible for you.

Things get really interesting when you start to combine information from multiple devices (and other systems) in novel ways. Now we're talking about the concept of *big data*—the analysis of disparate pieces of information not originally designed to be looked at together. Take a piece of data from this smart device, combine it with a piece of data from another smart device, and sometimes 1 plus 1 equals more than 2. The network puts all the pieces together and arrives at an interesting conclusion that leads to a unique action on yet another smart device. All these devices working in concert end up knowing more than any individual device knows—and, in many cases, more than you yourself know. It's a little scary, but holds a lot of promise for a more efficient, less labor-intense lifestyle.

 Note

The word "smart" gets thrown around a lot these days, especially in conjunction with the IoT. Walk down the aisles of your local Best Buy store and you'll find all manner of smart TVs, smart refrigerators, smart thermostats, and, of course, smartphones. This overuse (and misuse) of the term belies its descriptive value. Some of these so-called smart devices, such as smartphones, have incredible amounts of computing power built in. Others, such as smart TVs, don't have much if any built-in "smarts"; with these devices, the descriptor merely indicates that the device can connect to the Internet or other devices. I use the "smart" descriptor throughout this book because everybody and their brother (including most of the companies that produce IoT devices) uses it, but know that just because a device is called smart doesn't necessarily mean that it really is. With most IoT devices, the intelligence comes from the way the aggregated data from multiple devices is analyzed and acted upon, not from the devices themselves.

When Will the Internet of Things Arrive?

All this talk about the Internet of Things being the Internet of Tomorrow belies the fact that, in many ways, the IoT is here today. While certainly not in its most developed form, we do have an emerging IoT in the form of sensors, devices, cloud-based infrastructure, and data mining and analysis tools. Many of the pieces and parts are in place, just waiting for the final network to form.

Think about it. Today, more than a billion devices exist that contain embedded sensors, capable of capturing all sorts of data. All we need is for these devices to connect to the Internet, or to each other, for more intelligent purposes.

Go back to that example of the smart car. While it's true that today's cars don't have the capability of connecting autonomously to one another or to the Internet, they do contain lots of sensors that capture lots of information. Today, that information has to be manually retrieved when the guys at your auto repair shop hook their computers up to the car's little black box, but tomorrow all that connection and communication will happen automatically, behind the scenes. Again, many of the pieces are in place, just waiting to be connected in an intelligent fashion.

What is it exactly that we're waiting for? Some of it's technical, as the various companies that produce sensors and controllers and other devices sort out the necessary technologies and protocols that will enable all these devices to connect to and communicate with each other. Connecting everything together in the Internet of Things is technically complex, as you might imagine—and getting competing companies to talk with one another is several times more complex than getting IoT devices to do so.

Even when the technologies get in line, cost will be a factor, at least for a time; all those technological advancements cost money, and it takes time for economies of scale to be reached. In addition, the value proposition for connected devices must be established. How much more, after all, will you pay for a smart refrigerator than for a regular old dumb one?

It's likely, then, that the first large-scale adoption of IoT technology will come not in the consumer segment, but in the business and government markets. Businesses that can realize true cost efficiencies are most apt to embrace the enabling technologies. And city and state governments will find that the short-term investment in

IoT technologies will reap significant long-term rewards in the fields of infrastructure maintenance, traffic control, energy consumption, and the like.

That's all well and good, but let's face it, the most exciting implications of the IoT will come in the consumer space. (These are also the most potentially profitable applications, for those companies producing IoT devices and technology.) Think of everything that can be connected in a typical home—appliances, TVs and other electronic devices, heating/cooling systems, lights, plumbing and electrical systems, clothing, cars... the list of things for you to buy is staggering.

That said, few consumers are going to shell out big bucks to replace everything in their home with new smart versions overnight. The rollout will be slow and costly, most likely taking place as people replace old items that wear out with new smarter ones. Few people are going to dump their old appliances and systems just to implement smart tech; it will be an iterative process.

Expect, then, the IoT to have a long and involved gestation. It won't be fully in place next year or the year after that. We're talking a decade or several decades before the majority of items and systems are compatible with and connected to the IoT. But as more and more devices are connected, the benefits will start to build. And that's a good thing.

How Important Is the Internet of Things?

Most experts expect the Internet of Things to surpass the current Internet in terms of size, importance, and revenues. If all the predictions come true, it's going to be a very big thing indeed.

First, how many things will be in the Internet of Things? The Gartner research firm estimates that the IoT will connect close to 26 billion devices by 2020. Competing research firm Allied Business Intelligence (ABI) Research says the number will be more than 30 billion. Tech powerhouse Cisco prophesizes 50 billion devices in the same time frame; Nelson Research says 100 billion devices; Intel says 200 billion; International Data Corporation (IDC) says 212 billion. While it's clear that nobody really knows for sure, it's also clear that we're talking a lot of things being connected in the next decade and a half.

And all these things will generate a lot of revenue for a lot of companies. How much is anybody's guess. Gartner says that, by 2020, the economic impact of the IoT will be $1.9 trillion. IDC says the number will be more like $8.9 trillion. (What's a few trillion dollars between friends?)

Whatever the numbers end up being, it's clear that the IoT has the potential to radically transform everyday life, business, and the global economy as we know it.

There are some big changes up ahead—and a lot of people look to make a lot of money from them.

SMART CONNECTIVITY AND YOU

If everything plays out as predicted, the IoT will offer tremendous benefits to individuals, businesses, and larger entities. For consumers, these benefits range from the nice but relatively trivial (creating custom music playlists or television favorites lists for each member of your family, setting just the right temperature for your bath water) to the cost effective (minimizing energy usage throughout the day, running appliances and sprinkler systems when water usage is lowest) to the absolutely essential (automatically notifying authorities of home and personal emergencies, preventing on-the-road mishaps, scheduling home and auto maintenance).

Once the IoT takes hold, we'll all wonder how we got along without it. While it's true that the IoT will do a lot of things that we can already do ourselves (but often don't), in most cases it'll do them better and more reliably. The IoT promises to automate the tedious activities we all hate to do and, in the process, make all the things in our lives run better and longer.

In automating (and enhancing) the performance of day-to-day activities, the IoT concurrently promises to add more free time to our day. If we don't have to bother with all those tedious activities, we can spend that time doing something more productive or rewarding. Or maybe just watching more TV. In any instance, time spent doing menial chores can now be spent doing something different. That's something to look forward to.

Knowing that the IoT is in your future, what can you do to prepare for it? In many instances, not much; you'll simply have more IoT-enabled choices when you go to the store, until ultimately you will have only IoT-enabled items to choose from. But you can prepare by investing in IoT-ready architecture today. Make a point to buy "smart" electronic devices and appliances that are Internet-ready—that is, that can connect to the Internet. Make sure your home is Wi-Fi–friendly, or if you're building a new home, wire all the rooms for Ethernet. (Some IoT devices will work better with a faster, more reliable wired Internet connection.) Make sure your Internet service provider (ISP) offers plenty of bandwidth, and pay for more if necessary. Familiarize yourself with developments in the IoT space, and try out new technology as it becomes available.

And don't ignore the smart technology that's available today—even in its crude introductory forms. You can get your hands dirty with a variety of home automation devices, wearable technology, smart car features, and more IoT-based technology that's currently or soon to be on the market. We'll talk about all of these items—and more—throughout the book, so keep your eyes open to the IoT tech you can work with today.

For businesses, there are some very concrete things you can do to prepare for the IoT. Make sure your facilities are fully connected, and that you have the most possible bandwidth available on your network. Consider setting up a separate network just for the IoT, and have your IT staff determine what devices and services should reside on this second network. Make sure you have plenty of storage (and the capacity to add more) to handle all the data that is apt to be collected by IoT devices. Work to ensure that this data and your entire network have adequate security going forward. And do so without blocking access to the very devices necessary to make the IoT both effective and efficient. (Too many IT departments think that increased security means banning the use of outside devices on the corporate network; since the IoT is all about these devices, banning them pretty much defeats the purpose.)

The main thing is to keep an open mind and an open ear about the IoT. Whether in the home or at work, make sure you have the infrastructure necessary to make the IoT work. And be prepared to spend some money up front to set yourself up for the IoT—and then reap the rewards when everything finally starts working together.

2

Smart Technology: How the Internet of Things Works

The Internet of Things, as it appears to be developing, is a technological tour de force. It uses a variety of existing and soon-to-be-developed technologies and protocols, combines them in new and interesting ways, and ends up automating all manner of day-to-day activities and interactions. It puts together multiple, relatively simple technologies and ends up with something that is greater than the sum of its parts.

How exactly, then, does the Internet of Things work? Let's take a look.

Understanding the Internet of Things: The Big Picture

The Internet of Things combines a variety of technologies into a semi-autonomous network. Put simply, the IoT connects individual devices to the network and to each other. The network is also connected to software and services that analyze the data collected by the connected devices and use that data to make decisions and initiate actions from the same or other devices.

These devices typically connect to the network via a wireless technology. This wireless tech can be the now-ubiquitous Wi-Fi or some future purpose-driven wireless protocol. Some devices will connect via Bluetooth (a shorter-range technology) to a master device that then connects to the network via Wi-Fi.

As you can see, IoT's backbone is the network—or networks, plural. You'd think by the name (Internet of Things) that the network everything connects to would be the Internet, and it sometimes is. But IoT devices don't have to be connected to the Internet; instead, they can be connected to proprietary or purpose-built networks. In fact, IoT devices don't have to be connected to the Internet or an always-on network. Some devices can store collected data while not connected, then transmit that data when they connect at a later time.

For the data collected by these connected devices to be of value, there needs to be some sort of software or service to analyze, process, and act on. The software can reside on another device, on a separate controller, in a corporate data center, or in the cloud. These applications can work in an automated or semi-automated fashion or, less commonly, be controlled by real, live human beings.

So, in a nutshell, we're talking about:

- Devices that contain
    - Embedded sensors that capture or generate data
    - Wireless transmitters and receivers that connect to a larger network
- Network backbone, to which all devices ultimately connect via various wireless technologies
- Software applications that
    - Analyze and process all the collected data
    - Initiate appropriate actions

Figure 2.1 details how all of these technologies work together.

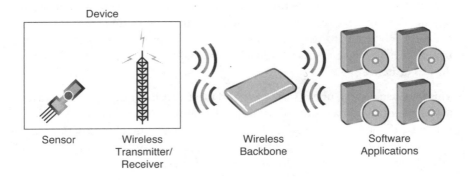

Device

Sensor   Wireless   Wireless   Software
              Transmitter/   Backbone   Applications
              Receiver

Figure 2.1   *The technologies of the Internet of Things.*

The Internet of Things that results from all these combined technologies isn't really a thing unto itself, but more of a concept that can be applied to all manner of situations and devices. That is, there is no one single overruling universal Internet of Things; instead, the concept behind the Internet of Things is appropriated by various programmers and companies to provide specific functionality.

In essence, each individual IoT application is purpose-built. We could be talking about the network of sensors, devices, and applications built into a smart car, a smart house, a smart factory, or even a smart city. Each application of the concept incorporates the same types of smaller devices connected to the same types of wireless networks, but each set of devices works toward its own unique end. Maybe one network of things will communicate with another network of things for an even larger goal, but that isn't necessary. What is important is that each purpose-built application does the job it needs to do, as enabled by the underlying IoT technologies.

Building the Internet of Things

Given all the technologies inherent in the Internet of Things, just how does the whole thing get built? It's a daunting process that presents both challenges and opportunities for those companies seeking to get into it.

Experts envision three stages of development before the full potential of the IoT is realized. It's all about installing and connecting the devices, enabling two or more devices to work together for a joined purpose, and then creating applications that analyze the connected data and initiate even more complex operations.

Stage One: Device Proliferation and Connection

Because the foundation of the IoT is built on a network of devices, it's not surprising that the first stage in constructing the IoT is all about getting more devices out there. We're talking devices of all types—sensors, processors, smart hubs, you name it. These can be freestanding devices or smaller devices embedded into larger devices or items.

This first stage is well under way. More and more consumer devices are connectable, from fitness trackers to televisions to thermostats. These devices connect wirelessly to the Internet, typically using Wi-Fi, and gain functionality from that connection. For that matter, smartphones and personal computers also function in this manner as devices, thus adding to the size of the underlying network.

 Note

Even though today's connected devices are typically called "smart" devices, they're not really all that smart. But they do deliver some basic types of functionality, and connect to the network to either send or receive data to/from other devices, which makes them a little bit more intelligent than dumb non-connected devices.

As we've discussed, there are various network technologies currently in place that can form the base of the necessary IoT connectivity. But there may be additional opportunities in the form of newer networking technologies—and, of course, somebody has to build, install, and connect all those billions of devices. We're just starting.

Stage Two: Making Things Work Together

The second stage of development is where we get two or more of these so-called smart objects to work together for some greater purpose. We're talking sharing data to automate some process. The data from one device is transmitted to a second device, which then makes some sort of decision and initiates a given operation.

For example, you might have an in-ground moisture sensor planted somewhere in your backyard. This sensor sends its collected data over the network (Bluetooth, Wi-Fi, or other) to another device embedded in the controller for your yard's sprinkler system. The sprinkler system is programmed to operate based on the data received from the moisture sensor device. If the yard is wet enough, the sprinkler system won't run; if the yard is dry enough, it will run a specified cycle; if the yard is even dryer, the sprinklers will run a longer cycle. You don't have to bother with anything, because the two connected devices do all the monitoring and decision-making for you.

This stage is about automating simple tasks and programming the necessary devices to do that. It's not hyper-intelligent; the devices perform operations entered into memory. It's Google-type, algorithm-driven decision-making. If A then B, if C then D, and so forth.

We're cautiously entering into this second stage, more in some industries than in others. The challenge is in determining all the boring daily activities that can bene-fit from this type of automation, installing the necessary sensors and other devices, and then programming those devices to do the simple tasks necessary to automate the underlying activities. It's a matter of magnitude; there's a lot of what we do that can be automated in this fashion.

Stage Three: Developing Intelligent Applications

Basic automation is like a decision tree: If the sensor says this, then do that; if the sensor says that, then do this. If we want to truly take advantage of the vast amounts of data collected from the IoT, we need applications that can act on larger, more complex, and often more obscure data sets. It's more than just tying together the behavior of two simple objects; it's about creating sophisticated inter-relationships that utilize and analyze additional data points.

Consider your home thermostat. A simple automation would involve an outside temperature sensor connected to the thermostat, which functions as a controller for your furnace and air conditioner. When the outside temperature rises to a cer-tain level, the thermostat is programmed to turn on the air conditioner. Useful, of course, but rather limited.

In the third stage of IoT development, this simple sensor/thermostat relationship will be augmented by other data, such as the day's weather forecast, historical temperature patterns, even historical patterns of room usage by the people in your house. An application (app) could then be developed that takes all this data (and more) and predicts exactly how long the air conditioner should run and when. It's not reactionary; it's predictive—and that's where the system gains intelligence.

There is much opportunity here, not just for the manufacturers of the devices and networking hardware, but also for app developers. There's a lot of work necessary to make this third stage a reality.

Understanding Smart Devices

When it comes to understanding how the Internet of Things works, it's best to start at ground level, with the many different devices that make up the front lines of the concept. Everything starts with these devices; the Internet of Things needs these things to exist.

What's a Thing?

There are many different types of devices that can be connected to the Internet of Things. A device can be something big and complex, like a car or a house. It can be something you use in everyday life, such as a golf club, or a printer, or a pair of sneakers. It can be something very, very small, such as a discrete sensor inside your car or house or golf club, a single part of a much larger and more complex device.

For that matter, what the IoT calls "things" don't have to be actual physical things. A thing can be a piece of data—status information such as your location or the room temperature—collected through a separate general-purpose device, such as a thermostat, smartphone, or computer. Put another way, the physical thing itself doesn't have to be in the IoT, although data about the thing must.

Know, though, that most IoT devices are simple sensors that monitor something happening nearby. These simple things either record or transmit the information they gather across the network to some other device or service.

Building Blocks

Larger things in the Internet of Things are actually collections of these smaller devices. For example, what we might call a smart car is actually the collection of all the small sensor devices built into the car and its parts, as illustrated in Figure 2.2. The individual sensors connect to an in-car network to transmit the data they collect to a master computer that runs the necessary controller software. This computer—actually, just another type of device—makes decisions based on the data it receives, and then sends instructions to other devices (controllers) also built into the car. The computer might also transmit some of the data it receives, or analysis based on that data, to other devices and services outside the car's environment—your auto repair shop, for example, or the car's manufacturer, or even the company you work for (if the car is supplied by the company, that is).

So big things are made out of little things, with more complex devices containing two or more embedded devices. It's like constructing something big out of a collection of smaller building blocks. And the bigger thing becomes "smart" when it communicates with other similar devices.

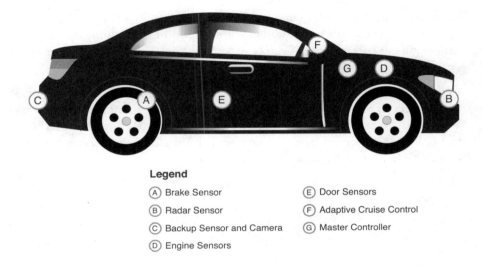

Legend

(A) Brake Sensor

(B) Radar Sensor

(C) Backup Sensor and Camera

(D) Engine Sensors

(E) Door Sensors

(F) Adaptive Cruise Control

(G) Master Controller

Figure 2.2   *Multiple small devices connected together to form a larger smart device.*

Deconstructing a Device

There's one component that's common across all types of devices. To connect to the IoT (and to other devices), a device must include some sort of radio that can send and receive data. This may be a Wi-Fi radio that connects directly to the nearest Wi-Fi network, much like your notebook computer connects to a Wi-Fi hotspot today. It may be a Bluetooth radio that only has enough power to connect to a larger device, such as a controller or computer. Or it can be a new type of low-power radio that operates on a wireless network built specifically for the transmission of this type of small and simple data.

Given the miniature size of most basic devices, the wireless transmitter/receiver needs to be as small as possible. In addition, the device itself needs to use as little power and network bandwidth as possible.

In terms of energy usage, the low power profile follows from the fact that these are simple, often single-purpose devices; it doesn't take a lot of electricity to collect and transmit small pieces of data. It's also somewhat of a requirement; if the device is a sensor in a piece of clothing or sporting equipment, it won't have much power to draw from, a small battery at best, and thus needs to draw as little power as possible.

In terms of bandwidth, most sensors transmit only limited types and amounts of data, and don't need much bandwidth to do so. A sensor transmitting the current temperature or how many steps you've walked has low bandwidth requirements by nature.

And, just to clarify, while most devices will connect to the IoT wirelessly, some will also connect via wires and cables. It might be easier to wire together the devices in a smart car, for example, than forcing wireless communications on the whole of them. What matters most is that the devices are connected; how they connect is a detail. (An important detail, mind you, but still a detail.)

 Note

The low power/low bandwidth nature of these devices points out how the IoT builds on simple devices (but lots of them) to assemble data that then is used to make more autonomous and intelligent decisions.

Store and Forward

There's something else that an IoT device needs to be functional. That something is *memory*.

When we say that data is transmitted from a dedicated device to another device or service, it doesn't mean that the data is transmitted immediately. Much collected data is not time-sensitive, which means it can be saved and transmitted when it's most convenient (or cost effective). It's also possible that a given device won't be constantly connected to a network; data may be captured when the device is offline but not transmitted until it is next online.

To do this, a device must have what is called *store and forward* capability. That is, the device must have some small amount of data storage, typically in the form of a tiny, solid-state memory chip. The data collected can then be stored in the device's memory until the next connection cycle, and then forwarded to wherever it needs to be forwarded.

So, in addition to an IoT device being small and low-powered and having some sort of wireless connectivity, it must also have built-in storage capacity.

Understanding Network Connections

For individual devices to communicate with other devices in the IoT, they must be connected to some sort of network. A network exists when two or more devices are connected together, typically for the purpose of transmitting or sharing data or other communications.

When we're talking about the IoT, the network connection is typically wireless. That's a matter of practicality; it would be difficult if not impossible to connect all the billions of smart devices to one another using cables, of whatever sort. It's a lot easier to connect devices to one another, to central hubs, and to the Internet when those connections are wireless.

How Traditional Networks Work

In a traditional network, wireless or wired, individual devices do not connect or communicate directly to each other. That is, the computer in your living room doesn't communicate directly with the one in your office, nor does your smartphone connect directly to your modem to access the Internet.

Instead, each and every device on your network connects directly to a central hub called a *router*. As you can see in Figure 2.3, all transmitted data passes through the router en route to another device on the network or to the Internet. (Assuming, that is, that the router itself is connected to the Internet, typically via some sort of modem.) Some call this a *hub-and-spoke* approach, and it's how things work in most home and office networks today.

Transferring Data Over a Network

Data transferred over the network is broken into smaller pieces, for easier transmittal. When you're sending a file to another device on a network or over the Internet, that complete file isn't sent at once. Instead, the file is broken into multiple smaller data packets, which enables large amounts of data to be transferred without clogging the connection. The data packets are then reassembled at the receiving end by the appropriate networked device, as shown in Figure 2.4.

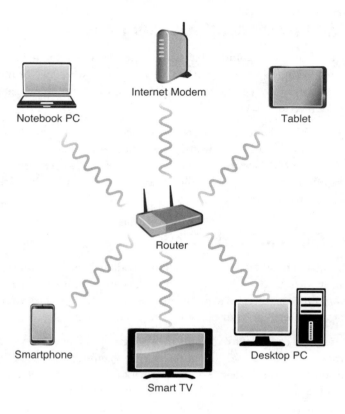

Figure 2.3   *Devices connected to a central hub in a traditional network.*

To enable this disassembly/transmittal/reassembly process, all networked hardware must work in tandem with a predescribed set of networking transfer protocols. These rules determine how data is transmitted across the network.

The de facto standard network protocol today, used for both Internet and local area network (LAN) connections, is called *TCP/IP* (that's short for *Transmission Control Protocol/Internet Protocol*). The *IP* part of this protocol provides the standard set of rules and specifications that enables the routing of data packets from one network to another. The *TCP* part of the protocol supports the necessary communication between two devices; it takes network information and translates it into a form that your network can understand. In other words, IP sets the rules and TCP interprets those rules.

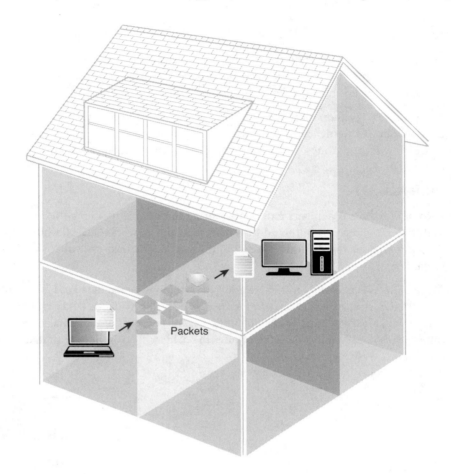

Figure 2.4 *Transmitting multiple data packets over a network.*

Here's how it works in practice. Let's say that you want to copy a file from the PC in your home office to the PC located in your basement. When you click the Copy button, TCP establishes a connection between the two computers, and then IP lays down the rules of communication and connects the ports of the two computers. Because TCP has prepared the data for transmittal, IP then takes the file, breaks it into smaller pieces (data packets), and puts a header on each packet to make sure it gets to where it's going. The TCP packet is also labeled with the kind of data it's carrying and how large the packet is.

Next, IP converts the packet into a standard format and sends it on its way from the first computer to the second. After the packet is received by the second PC, TCP translates the packet into its original format and combines the multiple packets back into a single file.

Understanding IP Addresses

For TCP/IP to work, each device on a network needs to be properly configured with the proper information. In particular, each device needs to be assigned a local *IP address*, which is how the device is known by the network.

An IP address is a numerical label, kind of like a street address but with all numbers. In today's Internet, an IP address is a 32-bit number expressed as a "dot address" with four groups (or *quads*) of numbers separated by periods or dots, like this:

192.106.126.193

Each of the decimal numbers represents a string of eight binary digits—0s and 1s, as it were. The first part of this address represents the network address, while the latter sections represent the address of the local device (also known as the *host address*).

An IP address is necessary for a network router to know which data goes to which connected device. As you can see in Figure 2.5, TCP/IP broadcasts data to the router, with a particular IP address identified as the recipient of that data. The router reads the IP address and then routes the data to the computer with that address.

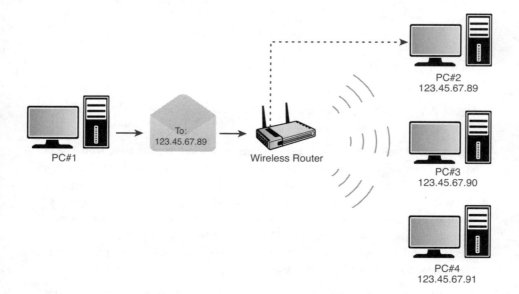

Figure 2.5   *Sending data to a specific network address.*

Every server or device connected to the Internet today is assigned its own IP address. In tomorrow's world of the IoT, every single device, no matter how small, must also be assigned an IP address.

This produces a bit of an issue, as the huge number of devices that need to be connected will easily exceed the available number of IP addresses—at least under the current IPv4 (that's version 4 of the Internet Protocol) address scheme. IPv4 allows for some 4.3 billion unique addresses, a large percentage of which are already assigned to existing devices.

The solution is the rapid adoption of the next-generation Internet Protocol, IPv6. (Yes, that's skipping from version 4 to version 6.) This new protocol expands the pool to a theoretical maximum 340 undecillion—that's 240 trillion trillion trillion—addresses, which should be more than enough to handle all the possible IoT devices. It's fair to say that the full-scale implementation of the IoT would not be possible without IPv6.

 Note

IPv6 expands the number of available IP addresses by moving from a 32-bit address (in IPv4) to a 128-bit address. It's like changing from a three-digit number to a twelve-digit number—there are simply more possible numerical combinations when you have more characters to work with.

Examining Wireless Technologies

In a strictly wired network, devices connect to the router via Ethernet cables. If the network is wireless, the router contains a small radio that transmits and receives the wireless signals from all connected devices.

There are several wireless technologies in use today. All transmit and receive (RF) signals at a specific frequency. These are the same types of signals used in AM and FM radio; the big difference is that an AM/FM radio only receives signals, whereas wireless networking devices both send and receive.

Understanding RF Technology

How exactly do RF signals work? It all starts with a single radio wave, which is nothing more than a pulse of electromagnetic energy. Radio waves are generated when a transmitter oscillates at a specific frequency. The faster the oscillation, the higher the frequency. An antenna is used to amplify and broadcast the radio signal over long distances.

To receive a radio signal, you need a radio receiver. The receiver is tuned to a specific frequency to receive signals oscillating at that rate. If the receiver is not tuned to that frequency, the radio waves pass by without being received.

 Note

> When we're talking about home networks, the Internet, and the Internet of Things, the radio in each connected device functions as both a transmitter and receiver.

RF transmissions are spread over a broad range of frequencies, which are measured in cycles per second. For example, 93.5MHz is a frequency of 93,500,000 cycles per second. (MHz is shorthand for megahertz, or millions of cycles per second; GHz is shorthand for gigahertz, or billions of cycles per second.)

Current wireless networks use two distinct RF frequencies. Earlier equipment works in the 2.4GHz band (frequencies between 2.4GHz and 2.48GHz), while newer equipment can also utilize the 5GHz band (frequencies between 5.15GHz and 5.85GHz).

The 2.4GHz band is free for anyone to use, for any purpose. That's both good and bad—good because it can be used at no cost (without potentially expensive licensing fees), but bad because space within the band is finite, and several other types of devices also use this band.

 Note

> The 2.4GHz frequency range is alternately called the ISM band—for instrumentation, scientific, and medical usage.

Currently, the 2.4GHz band is used by 802.11 Wi-Fi networks, Bluetooth networks, newer cordless telephones, newer models of baby monitors and garage door openers, microwave ovens (!), urban and suburban wireless communications systems (including many emergency radios), and some local government communications in Spain, France, and Japan.

The 5GHz band, on the other hand, is relatively unused—and a lot wider, with more frequencies that can be used. (It stretches from 5.15GHz to 5.85GHz, remember.) Some cordless phones use this band, but not too many, so there's not a lot of competition for frequencies. Like the 2.4GHz band, it's unregulated, which means that it's free for any device to use—which becomes important when we start talking about the IoT.

Wi-Fi

Most home, business, and public wireless networks today utilize the Wi-Fi protocol. Wi-Fi (short for *Wireless Fidelity*) is the consumer-friendly name for the IEEE 802.11 wireless networking standard. Most of today's wireless networks are technically Wi-Fi networks and use Wi-Fi–certified products.

 Note

The IEEE is the Institute of Electrical and Electronics Engineers, and it does things like ratify different technology standards. In the case of Wi-Fi, the technology is now regulated by a subgroup called the Wi-Fi Alliance. Learn more about it at the Wi-Fi Alliance website (www.wi-fi.org).

But, here's the thing. There isn't a single Wi-Fi protocol. Instead, there are multiple 802.11 protocols, each designated by a one- or two-letter suffix. Different versions of Wi-Fi offer different levels of performance; the latest version, 802.11ac, operates on the lesser-used 5GHz band (but is backwards compatible with older devices operating on the more-established 2.4GHz band) with enough range to cover a fairly large house.

Wi-Fi networks utilize the hub-and-spoke configuration common to traditional networks. Each Wi-Fi–enabled device connects to the central hub or router and, via the router, is connected to other devices that are also connected to the router. The notebook computer in your living room does not connect directly to the streaming media box in your bedroom; the signals from one device go first to the hub and are then routed to the other device. If you have a dozen wireless devices connected to your Wi-Fi network, that's a dozen devices connected directly to the router.

Each of these devices has to be manually configured to connect to the network, typically employing some sort of wireless security technology. That's a lot of initial set-up necessary; while subsequent connections can be automated, every single Wi-Fi device must be connected by hand, so to speak, on first connection. If you were to have a hundred small IoT devices scattered throughout your living quarters, connecting them all via Wi-Fi would be impractical.

That said, given the ubiquitous nature of Wi-Fi today, it would be convenient for IoT devices to utilize Wi-Fi technology. And, in fact, many current devices do. If you have a smart TV in your living room, for example, it connects to the Internet via Wi-Fi, as do most other current devices.

But it's unclear whether Wi-Fi is the best solution for all the connectivity required to implement a global IoT in the future. To that end, other wireless protocols might be better suited to handle the wireless communications between small sensor devices; these devices might then connect to a master device that then connects to the Internet or an appropriate service via Wi-Fi.

Bluetooth and Bluetooth Smart

When it comes to close-quarters, device-to-device wireless communications, *Bluetooth* is an interesting option. Bluetooth is similar to Wi-Fi in that it's a wireless technology that operates via RF transmission in the 2.4GHz frequency range. But it's different in that it isn't intended for use in hub-and-spoke networks; instead, it's designed for direct communication between devices—what's called *peer-to-peer networking*—which is how many speculate the IoT will end up working.

Today, Bluetooth is used to connect various devices to one another over short distances, such as your smartphone to your car audio system, or a wireless mouse to your desktop computer. It is not a good technology for transmitting large data packets over long distances; for that, Wi-Fi is still king.

Bluetooth radios are extremely small, much smaller than Wi-Fi radios, which makes the technology ideal for miniature IoT sensor devices. They also consume very little power, which also fits into the IoT profile.

When one Bluetooth device senses another Bluetooth device (within about a 30-foot range), they automatically set up a connection between themselves—once the initial manual configuration is complete, of course. This connection is called a *piconet*, which is a kind of mini-network—a *personal area network* (PAN), to be specific.

In a piconet, one Bluetooth device is assigned the role of master, while the other device—and any subsequent devices, up to eight in total—is assigned the role of slave. The master device controls the communication, including any necessary transfer of data between the devices. Each piconet can contain up to eight different devices.

This means that, over short distances, a device such as a smartphone can connect to, synchronize with, and even control the other electronic devices in your home, office, or car—such as your personal computer, printer, television set, home alarm system, or car audio system. All this communication takes place in an ad hoc fashion, without your being aware, totally automatically.

Here's an example of how this works today. You have a smartphone that contains your contacts list. You need to synchronize this contacts list, which includes phone numbers, with your car's built-in dialing system. To do so, all you have to do is carry your smartphone with you when you enter your car. When you're close enough (and your car is powered on), your phone automatically connects to your car's system via Bluetooth, and then automatically synchs the contact data between your phone and the car. If you've added a new contact to your phone, it's also added to your car's system. You can then dial any of your contacts from your car's dashboard—all the communication is synchronized via the Bluetooth connection; no manual intervention necessary on your part.

If that doesn't sound IoT-friendly enough, there's *Bluetooth Smart*, a variation on Bluetooth technology developed especially for the Internet of Things. Bluetooth Smart is a version of Bluetooth that consumes only a fraction of the power of traditional Bluetooth radios. This power efficiency makes Bluetooth wireless connectivity practical for devices that are powered by small coin-cell batteries.

In addition, Bluetooth Smart operation can be tweaked by the hosting app. Instead of being limited to Bluetooth's standard 30-foot range, a Bluetooth Smart device can be optimized to work up to 200 feet away—ideal for in-home sensors where a longer range is necessary. Naturally, extending the range uses more power, so devices that don't need the range can be configured to use less power instead. It's an app-by-app tradeoff, perfect for the IoT.

Cellular Networks

Some IoT connections will utilize cellular networks—the same networks you use to connect your mobile phone. For example, the devices in a smart car might communicate with your home network or to the auto dealership by literally dialing in like a cell phone. It's quick and easy and utilizes a somewhat ubiquitous existing technology.

Cellular technology works a little differently than either Wi-Fi or Bluetooth technologies. While cellular signals carrying voice, text, and digital data are transmitted via radio waves, this information is transmitted not to a central hub in a small network of devices (as it is with Wi-Fi) or even directly from device to device (as it is with Bluetooth), but through a global network of transmitters and receivers.

These networks of transmitting/receiving towers are built on a cellular design. (Hence, the terms "cellular network" and "cellular phone.") That is, a mobile phone network is divided into thousands of overlapping geographic areas, or *cells*. A typical cellular network can be envisioned as a mesh of hexagonal cells, as shown in Figure 2.6, each with its own *base station* at the center. The cells slightly overlap at the edges to ensure that users always remain within range of a base station. (You don't want a dropped call when you're driving between base stations.)

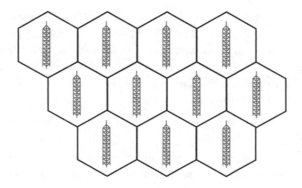

Figure 2.6   *Cells in a cellular network.*

 Note

> The cells in a cellular network vary in size, depending on how many calls
> are conducted within that geographic area. The smallest cells, which might
> cover only a few city blocks, are those where there's the heaviest popula-
> tion density, and thus the largest demand for service. The largest cells are
> most often in rural areas with a smaller population per square mile.

The base station at the center of each group of cells functions as the hub for those
cells—not of the entire network, but of that individual piece of the network. RF
signals are transmitted by an individual phone and received by the base station,
where they are then re-transmitted from the base station to another mobile phone.
Transmitting and receiving are done over two slightly different frequencies.

Base stations are connected to one another via central switching centers which
track calls and transfer them from one base station to another as callers move
between cells; the handoff is (ideally) seamless and unnoticeable. Each base station
is also connected to the main telephone network, and can thus relay mobile calls to
landline phones.

Given the capability of transmitting data over long distances, cellular networks
may play some part in IoT connectivity. A cellular connection wouldn't be the best
choice for connecting two devices over a short distance (such as inside a car), but
could function as the primary connection between a master device and a more distant IoT hub or service.

Mesh Networks

As billions of new devices join the IoT, the networks they connect to will become increasingly crowded, leaving those networks—and the Internet in general—unable to handle the increased traffic. This is leading some companies to develop their own specialized networks for IoT connectivity.

One option is to employ a purpose-built, small-scale wireless network that enables devices to connect directly with each other, one after another, kind of like runners passing a baton in a relay race. This type of network, called a *mesh network*, enables the automatic hand-off of wireless signals from one device to another.

A mesh network is the opposite of a traditional centralized network. Where a traditional network connects all devices to a central server or hub, a mesh network can theoretically connect devices from one end of a city to another, one after another, as shown in Figure 2.7. Each individual device on a mesh network might only have a range of 30 feet to 300 feet, but when they're connected end-to-end, they can cover a wide area.

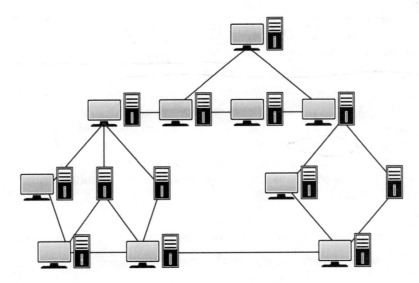

Figure 2.7   *Devices connected one after another in a mesh network.*

A mesh network can contain thousands of individual devices all operating in concert. Because there are multiple routes through a mesh network, the failure of any individual device won't bring down the entire network; the signals will find a way using connections to alternate devices.

There are several protocols in development for IoT mesh networks. These protocols are being developed by INSTEON, theZ-Wave Alliance, and theZigBee Alliance. The INSTEON and Z-Wave protocols are proprietary, where ZigBee offers certification to ensure adherence to their standards. At present, networks based on different protocols are not interoperable, although it's possible to connect two mesh networks together using hubs.

Proprietary Cellular Networks

Other companies are developing their own purpose-built wireless technologies for IoT connectivity. Like the mesh networks just discussed, these machine-to-machine (M2M) networks are designed specifically for connecting small devices, not people or computers. By offloading M2M communications onto these dedicated networks, the current Internet will maintain bandwidth for effective operations.

Probably the most visible at this stage of the game is a French company called-Sigfox, which is employing its own version of a World War I-era radio technology that submarines used to communicate with each other under water. This technology enables very small bits of data—think Morse code messages—to travel relatively long distances. It operates in the 900MHz frequency range, the same band used by some cordless phones and baby monitors today.

Sigfox is using this technology to build a dedicated low-power, low-bandwidth cellular network optimized for the types of short data transmissions common with IoT devices. The network isn't the fastest around, transmitting small bits of information at just 100 bps, but typical sensor data doesn't need more speed than that. Consider the data collected and sent by a traffic sensor or water meter; this type of device typically sends only a few packets of data on an intermittent basis. Speed isn't important.

What is important is the ability to connect lots of these little devices. Sigfox trades speed for capacity; to that end, its network can support millions of connections.

In addition, by using the 900MHz band instead of the higher frequencies used by today's mobile carriers, Sigfox can space its towers further apart than with typical cellular networks. Even better, it's a low-power solution, so that the power it takes to transmit a given packet of data is a fraction of what is necessary with a tradiional cellular radio.

The company is currently building networks in the United Kingdom (UK), Netherlands, Russia, and Spain, with its first United States (US)-based network in San Francisco. At this time, Sigfox is specializing in industrial networking. In Spain, it is employing its network to connect several million home security systems. In France, Sigfox's networks are being used to connect water meters, electronic billboards, and monitoring devices for seniors.

 Note

Interestingly, Sigfox's low-power networks are relatively inexpensive to build out. The company's Spanish network cost less than $19 million and took only seven months to complete. wow! That's cheap

Sigfox isn't the only company investing in proprietary networking technology. For example, Link Labs builds long-range M2M networks for public and private companies; On-Ramp Wireless builds custom networks for oil field operators; and Iotera is building a network to monitor global positioning system (GPS) trackers for pets and children.

Which Technologies Are Best?

Not everyone agrees with Sigfox that a separate network is needed for the IoT. Some view the development of proprietary networks as unnecessary and unproductive, perhaps enough so to hinder the development of the IoT. If competing proprietary or regional networks are developed, network cross-compatibility becomes an issue and we may end up with the same sort of splintered market that plagues today's cellular networks in the United States.

The thinking of some is that today's existing "big three" wireless technologies are sufficient to handle the needs of the IoT. Cellular will be used for wide area networking, Wi-Fi for local area networking, and Bluetooth for personal area networking. For companies developing IoT devices, using one of these existing technologies is relatively easy—potentially a lot easier than working with a newer, less ubiquitous technology.

That said, there's nothing to stop companies from developing their own proprietary connection solutions within larger smart devices or areas. Does it really matter whether the sensor devices in your car communicate via Bluetooth, Sigfox, or some new type of mesh network? At the end of the day, you only care that they *do* communicate with each, and then send their information from your car to your house or dealer or wherever, using technologies that are easy for you to work with.

Understanding the Data

The data collected by a given device is specific to that device's stated use. That is, different devices will collect different types of data.

For example, a sensor in a water meter collects data about a home's water usage—how much is used when. A sensor in a home thermostat collects data about when the furnace and air conditioning (A/C) are used, the temperature of the house (and the outside temperature) at different times of the day, and so on. Sensors in your car collect information about the engine temperature, oil level, and so forth. A sensor embedded in a highway collects data about the amount of traffic passing by, and perhaps the ambient air temperature.

The data collected by a given sensor is then transmitted to another device or service that then compares the data to other data and makes decisions based on that comparison. The thermostat data, for example, could be used to determine when to automatically turn up or down the furnace or A/C, based on past usage. It could also be used in conjunction with energy rate data to turn up the A/C when energy rates are highest, or when the power grid becomes overloaded.

You can also get creative with the data you collect. Take the sensor embedded in your local highway. With a little modification, the same device that senses vehicle traffic can also be used to detect the presence of certain chemicals. This data can then be fed into a national homeland security database and used to alert the authorities when certain types of chemicals are detected, thus giving early warning to potential terrorist attacks.

Understanding Intelligent Applications

For data to be useful, it must be capable of being acted upon. This can be done manually, by humans who analyze the data and then make corresponding decisions. But building a people-centric process goes against current technology trends, where every action is triggered by some sort of algorithm. (Think Google.) For the IoT to work, it cannot be labor-intensive; it must be more automated.

The key is to create intelligent apps that are capable of reading data and then acting autonomously on that data, based on certain preset parameters. For example, an app connected to your dishwasher or clothes washer could analyze data collected from the sensor on your water meter and automatically initiate washing when water usage is below a certain level—or delay washing when usage is too high. Another app might collect usage patterns from motion detectors throughout your home, be able to predict when you're in a given room, and have the lights and heat adjusted to your preferences beforehand.

These intelligent apps will be usage-specific, of course. The apps you use in your house or car will be much different from those used in your hospital, fast food restaurant, or warehouse. But they're all necessary, or all you have is a bunch of useless data. For the data to become useful, you need the right apps.

Understanding Big Data

The data collected in the IoT becomes even more useful when data from different types of devices are combined in creative ways. It's a matter of dealing with what the techies call *big data*. This is simply a term for large amounts of data—data sets so large that they can't be managed with traditional relational database technology. For the IoT to be truly valuable, processes need to be developed that sift through these huge amounts of data to make the connections and correlations that result in intelligent decision-making. It's all about connecting the data collected by this sensor here and that sensor there, and coming up with a conclusion that wouldn't have been possible otherwise.

There are actually three challenges involved in processing all the big data gathered via the IoT. First, there's harvesting the data. Second, there's storing the data. And third, there's analyzing the data.

Data Harvesting

Data harvesting (sometimes called *data ingestion*) is a multi-step process that involves data collection by individual devices, and then transmitting that data to some sort of central database. It's all about the devices and the networking—and the database, of course. We've already covered most of what's involved here.

Data Storage

Data storage appears simple—deceptively so. All you need is a bunch of servers, probably cloud-based, with enough capacity to hold all the data collected. Sounds simple enough, especially with the continuing drop in the cost of storage.

It's really not that simple, of course, even if it is kind of an old school problem. Lots of companies get hung up on the storage part of things and never get around to the more important analysis component.

That's too bad, because there are lots of companies out there that can handle the database storage needs and several different approaches to take.

One popular approach is to go with a company offering *database as a service* (DBaaS) functionality, typically in the form of cloud data warehousing. There are lots of options here, including Amazon Redshift, Enterprise Hadoop from Hortonworks, and Cloudera Enterprise. These database management and automation services alleviate the need for companies to install, manage, and operate their own large databases—freeing up valuable resources for the more important data analysis phase.

Similar to DBaaS providers but with even more functionality are the services offered by *managed service providers* (MSPs), such as All Covered and Treasure Data. These companies let you outsource not just data collection and storage but also basic analytics, typically in the form of extracting specified information from the main data. With an MSP doing the heavy lifting, a company can then focus its attention on detailed data analysis—and acting on that analysis.

Data Analysis

It's the third part of the challenge that's the most challenging. Assuming a company can manage or outsource the data harvesting and storage, there now comes the issue of how to extract value from the massive amounts of data collected. In other words, what does a company do with those massive amounts of data it has collected?

To work with data of this magnitude requires the development of apps that analyze the collected information for trends, patterns, and pressure points. It's a huge computing challenge, especially if you want results in something approaching real time.

When dealing with data on this magnitude, often collected (and thus stored) in an unstructured format, one of the major issues is making sure that you don't inadvertently skip over the important stuff while spending too much time on data that isn't important at all. It's a matter of separating the wheat from the chaff in regards to a particular app or operation.

 Note

Because of all the coming IoT data that will need to be analyzed, human resources folks predict a huge upsurge in the demand for data analytics experts. It's a good profession to get into.

But just analyzing the data isn't enough. For a company to truly take advantage of and benefit from this huge potential stream of real-time data, a company must develop a culture of data-driven decision-making. That is, companies need to go where the data leads them—not necessarily where old-line management might think they need to go. It's a brave new world, driven by all sorts of new data collected over the Internet of Things. Some companies will thrive on it; others won't.

Profiting from the Internet of Things

It's one thing to look at what the IoT can do for us as consumers. But the business world looks at the IoT from the other side of the table, as a vast opportunity to make money. That's what capitalism is all about after all.

How big an opportunity is the IoT? In Chapter 1, "Smart Connectivity: Welcome to the Internet of Things," we tossed around numbers that ranged from $1.9 trillion to $8.9 trillion by 2020. Even taking the low estimate, that's a lot of money to be made by somebody.

Not surprisingly, lots of big companies—including Cisco, IBM, Intel, Qualcomm, and Samsung—are betting big on building out the infrastructure behind the IoT. These companies see money to be made in selling the necessary hardware, of course, but also in providing additional services once everything is connected.

There are also multitudinous opportunities in individual industries that adopt the IoT. For example, appliance manufacturers can sell higher-priced, network-enabled refrigerators, dishwashers, and laundry equipment, as well as possible add-on services that enable all these appliances to work together. Suppliers to the auto industry will have a new class of parts to sell as auto manufacturers incorporate smarter devices into their cars. Warehouse operators will invest in IoT-enabled tracking and shipping systems that ideally will cut their current labor costs. (And some company will be selling the warehouses all the necessary IoT-enabled machinery and systems, too.)

So who benefits from the IoT? Communications companies, networking companies, tech companies of all stripes—including chip manufacturers, hardware manufacturers, cloud storage services, and app developers. Plus a ton of industry-specific players, of course, with a focus on equipment suppliers. And bet on today's big tech players—Apple, Google, and Microsoft—nosing their way into the IoT market as well.

 Note

Given how many hundreds of millions of devices will be battery-powered, it's a sure bet that battery suppliers stand to profit from the IoT in a big way.

While the big boys will obviously try to take a big share of the nascent IoT market, there should also be plenty of opportunity for new and smaller players to gain a presence. There will be lots of opportunities for everyone, or at least for those players offering unique benefits.

Resources

We've discussed quite a few companies and organizations in this chapter that are looking to help build—and profit from—the Internet of Things. To learn more about these companies, investigate their websites:

- All Covered (www.allcovered.com)
- Amazon Redshift (aws.amazon.com/redshift/)
- Apple (www.apple.com)
- Bluetooth Special Interest Group (www.bluetooth.org)
- Cisco (www.cisco.com)
- Cloudera Enterprise (www.cloudera.com)
- Enterprise Hadoop (www.hortonworks.com)
- Google (www.google.com)
- IBM (www.ibm.com)
- INSTEON (www.insteon.com)
- Intel (www.intel.com)
- Iotera (www.iotera.com)
- Link Labs (www.link-labs.com)
- Microsoft (www.microsoft.com)
- On-Ramp Wireless (www.onrampwireless.com)
- Qualcomm (www.qualcomm.com)
- Samsung (www.samsung.com)
- Sigfox (www.sigfox.com)
- Treasure Data (www.treasuredata.com)
- Wi-Fi Alliance (www.wi-fi.org)
- ZigBee Alliance (www.zigbee.org)
- Z-Wave Alliance (www.z-wavealliance.org)

SMART TECHNOLOGY AND YOU

When it comes to building the Internet of Things, you don't have to do much to prepare. The Ciscos, IBMs, and Intels of the world will be doing that work for you. All they ask is that you generously support their products and services in the years to come.

That said, you need to be aware of the changing infrastructure driven by the development of the IoT. For example, as you add more and more connected devices to your home, your home network might need more capacity. (After all, you don't want Netflix to stutter on your smart TV because your smart refrigerator is taking inventory at the same time.) That might mean upgrading to a newer wireless router or even paying for faster service from your ISP.

And you'll certainly be prompted (and tempted) to invest in newer "smart" electronics and appliances with enhanced IoT-like functionality. Some of these features might be worth paying for, some not. (And some might be duplicative; do you really need a smart TV, smart Blu-ray player, and smart media player that all offer the exact same Netflix functionality?)

So there's some decision-making you'll need to make in terms of what you buy and when. Do you replace your current "dumb" equipment with smart equipment now, or wait until something wears out? Do you invest in the smart versions of things now or figure things will only get smarter in years to come? Do you really want to spend a thousand dollars or more to replace all your light bulbs with smart light-emitting diode (LED) light bulbs, or just get by with cheap incandescents as long as they're available?

More important is developing the necessary mindset to accept the changes that the IoT is bound to bring. Let's face it, if the IoT only does half of what everybody's predicting, the impact on your daily life will be significant.

Think about it. Right now you're accustomed to setting your own thermostat, turning on your own lights, making your own coffee. If the day gets too hot, you turn up the A/C. When it gets dark, you turn up the lights. When you need caffeine, you brew a cup.

With the Internet of Things, however, you won't have to do any of these things. Your house will always be the perfect temperature, no matter the heat of the day or what room you're in. Your lights will turn on when you need them and turn off when you leave the room. Your coffee will be ready for you when you wake up in the morning, or when you normally take a mid-day coffee break.

In addition, your lawn will be properly watered and, with the inevitable adoption of the self-driving lawnmower, appropriately groomed. You'll no longer have to deal with handwritten grocery lists; when you run low on mayonnaise or eggs or beer, your smart refrigerator will know it and submit the list directly to your grocery store for automatic delivery. (Probably by drone. Definitely by drone.) And you won't need to manually program your digital video recorder (DVR), because your smart TV will know from experience what you like to watch and when, and queue it up for you.

That's a lot of decisions that you now make that you won't have to make in the future. Will you be comfortable ceding this much control to devices and apps? Will you trust the IoT to make these decisions for you?

And what will you do with all the time (and brain power) you currently spend making these decisions? One of the big potential benefits of the IoT is that you'll be rescued from all this old-fashioned manual thinking and labor. If you're not deciding what to watch tonight or writing out a grocery list or watering your lawn, what will you do instead? There's more leisure time in store, whether you want it or not.

In addition, some of us garner some small amount of satisfaction from performing these menial tasks. With the IoT doing everything for us, will our self-esteem suffer because we can no longer claim these minor accomplishments?

Perhaps it's a matter of deciding what to automate and what not to. Maybe you want to manually cut the grass or make out your grocery list, IoT be damned. There's something to be said for maintaining some degree of control over your life. The devices, systems, and apps don't have to do everything. As my two-year-old granddaughter insists on saying, "I do it." Perhaps there's a little two-year-old in all of us, and we want to maintain our self-sufficiency in the face of possible automation. Or maybe we're a bunch of lazy slugs who don't want to do anything ourselves. Hard to say until all the options are available

3

Smart TVs: Viewing in a Connected World

For many people, today's so-called "smart TVs" represent the first foray into the connected world of the Internet of Things. Just what is a smart TV, and how smart is it, really?

Whether or not today's smart TVs are truly part of the Internet of Things is an open question, but there's no question that these connected viewing devices are changing the way people watch TV and movies. Read on to learn more.

What Exactly Is Smart TV?

Let's be honest. "Smart TV," as the term is used today, is nothing more than marketing hype. The appellation refers to television sets or set-top boxes that offer connectivity to the Internet, typically via Wi-Fi wireless technology, as well as built-in Web 2.0 apps that enable viewing of various streaming video services, such as Netflix and Hulu. There's nothing inherently smart about a smart TV; it's a marketing term used to convey the ability to view Internet-based programming.

The concept of the smart TV isn't particularly new. Smart TVs have been around since 2007 or so, under many different labels, including "connected" TV, "hybrid" TV, "IPTV," and "Internet" TV. (One could even argue that the concept has actually been around since 1995's WebTV box, which served as an Internet client for traditional TVs.)

Note that a smart TV doesn't actually have to be a TV. Streaming media boxes and dongles that connect to a TV and offer the requisite streaming video connectivity also fit under the broad category of smart TV devices. So Roku and Apple TV set-top boxes are smart TV devices, as are the Google Chromecast, Roku Streaming Stick, and Amazon Fire Stick. For that matter, Blu-ray players and videogame consoles that offer streaming video connectivity are also classified as smart TV devices.

What's Inside a Smart TV?

At its most basic, a smart TV is a television set that can connect to and interact with the Internet. In practical terms, that means the television must include the following:

- Wi-Fi radio or Ethernet connection, for connecting to your home network.
- Central processing unit (CPU), the computer brain that manages all the device's operations and commands.
- Operating system (OS) that serves as the interface between the CPU and software-based applications.
- Graphical user interface (GUI) for displaying menus and other options.
- Software-based apps that enable connection to various web-based services. For example, a smart TV might have built-in apps for Netflix, Hulu, and Pandora. Most smart TVs come with several apps pre-installed; some smart TVs enable additional apps to be installed after purchase.

Some smart TVs also include apps and associated technologies that enable the device to play back media stored on your home network. In some cases, this capability is built into the OS, as with the Apple TV; in other cases, this capability is enabled by DNLA or UPnP compatibility.

 Note

DNLA stands for Digital Network Living Alliance, an industry trade group that promotes interoperability between different devices. In practical use, the DNLA specification indicates that a device is capable of playing digital media (video, music, and photos) from computers and other devices connected to the same network. UPnP stands for Universal Plug and Play and is a set of networking protocols that enabled connected devices to discover each other's presence on a network.

Some smart TVs include a built-in camera and microphone, like the one shown in Figure 3.1, for connecting with video-sharing and chat services, such as Skype. Some more advanced smart TVs use the built-in camera/microphone to navigate the onscreen menus, via a series of hand gestures or voice commands.

Figure 3.1   *The integrated camera on a Samsung smart TV.*

Naturally, a smart television set (not a set-top box) will also include a traditional television tuner for viewing broadcast, cable, or satellite programming. You typically switch from the normal viewing screen to a GUI menu for the web-based services and apps.

All smart TVs are controlled by some sort of remote control. Some remotes are basic affairs, with just enough buttons to navigate the onscreen menus. Others include keyboards (useful for typing in search terms), trackpads, even game controllers. Most smart TVs can be controlled by universal remotes, such as those in the Logitech Harmony line. Some smart TVs can be controlled by smartphone or tablet apps.

Remember, too, that a smart TV doesn't have to be a literal TV. A smart TV device, like the aforementioned Roku box, contains the same circuitry and apps as a literal smart TV, but without the TV part. Instead, the set-top box connects to a regular TV (typically via high-definition multimedia interface [HDMI]), enabling the TV to display media played on the external device.

What You Need to Use a Smart TV

Right out of the box, a smart TV has little or no functionality. To utilize all the features of a smart TV, you need to provide the following:

- An Internet connection.
- A home network that interfaces with your Internet connection. This can be a wireless (Wi-Fi) or wired (Ethernet) network.
- Electricity. Duh.

If you have a smart TV set-top box, you'll also need an HDMI cable to connect the device to your traditional television set.

What a Smart TV Does

So a smart TV is a TV or set-top box that integrates Internet capabilities. What exactly does that mean?

Most smart TVs can perform the following functions:

- Connect to the Internet via a local network. That means connecting to your home network and sharing your Internet connection. Most smart TVs connect via Wi-Fi, although some can connect via Ethernet.
- Play video content from web-based streaming video services, such as Netflix, Hulu Plus, and Amazon Instant Video.
- Play music from web-based streaming audio services, such as Pandora and Spotify.
- Play digital media stored on other devices connected to your home network.
- Access selected websites and web-based services, such as Facebook, Twitter, and AccuWeather. Some smart TVs offer full-fledged web browsers, although it's more common to find discrete apps for specific sites and services.

Note

Not everyone who owns a current-generation smart TV is actually using the "smart" aspects of the TV. According to NPD In-Stat, about 25 million U.S. households own a smart TV of one sort or another, but only about half of these homes (12 million) have their sets connected to the Internet. In other words, they're using their smart TVs just as TVs, nothing more. Unused functionality, it is.

Considering Smart TV Operating Systems

All smart TVs and smart TV devices are like mini computers, in that they include a built-in OS and the appropriate software or middleware to run the necessary apps. Now, these devices don't run a full-blown consumer OS, such as Windows, but rather smaller, more stripped down OS's developed specifically for these purposes.

 Note

Middleware is a layer of software on a device that acts as a bridge between the OS and the main apps.

There are a number of smart TV OS's in use today, many proprietary to a specific company or device. These include the following:

- Android TV, used in the Google Chromecast and selected Sony smart TVs
- Fire OS, used in Amazon's streaming devices
- Firefox OS, used on Panasonic devices
- iOS, Apple's mobile OS used in the Apple TV box (and iPhones and iPads, of course)
- Roku OS, used by Roku
- Tizen, a Linux-based OS used by Samsung
- webOS, a Linux derivative used by LG

 Note

webOS has an interesting history. It's a Linux kernel-based OS initially developed by Palm back in 2009 as a successor to their once-popular Palm OS platform. Hewlett Packard (HP) acquired Palm in 2010, and webOS was considered one of the key assets in that transaction; HP intended to use the OS in a variety of new products, including smartphones, tablets, and printers. That didn't really pan out, and by the end of 2011, HP had halted all webOS development. In 2013, HP sold webOS to LG Electronics, which uses it as the company's primary smart TV operating system.

This proliferation of OS's means that no two brands of smart TVs look or work exactly alike. While all these OS's do pretty much the same thing, they do it all differently; every company puts its own spin on onscreen menus, navigation, and operation. For this reason, you want to spend some time with a given interface when you're shopping for a smart TV or device.

Examining a Typical Smart TV

Most of today's smart TVs offer similar features and functionality. In addition to the normal TV features (screen, tuner, remote control, and so on), you get the Wi-Fi or Ethernet connectivity, onscreen GUI menus, and built-in apps that are part and parcel of the "smart" experience. Naturally, the onscreen menus and included apps differ from manufacturer to manufacturer and model to model, but all offer the same general approach.

Let us take, for our example, a typical higher-end smart TV, as of late 2014. We'll look at the Samsung UN50H6350, shown in Figure 3.2, a 50" diagonal LED-LCD model that sells for a little under $1,000. This model has all the bells and whistles that you'd expect from a TV in this price range, including smart TV functionality in the form of what Samsung calls its Smart Hub. It also includes a built-in camera and microphone, for live social networking and video chatting.

Figure 3.2   *Samsung's UN50H6350 smart TV.*

Before you can access the Smart Hub, you first have to connect the TV to your home network. This particular model includes both wireless and wired connectivity, so there is an Ethernet connection on the back if you want to use it.

 Note

If you have the option (and a convenient Ethernet cable), connecting a smart TV via Ethernet is a better option than using Wi-Fi. A wired connection is not only more reliable than a wireless one (you don't have to deal with weak or flakey Wi-Fi signals), but also faster—which is a godsend when you're watching high definition (HD) streaming video.

Assuming that you'll be connected via Wi-Fi, like the vast majority of users do (it's just easier), you have to configure the TV to recognize and connect to your home network. You do this from the Network Settings setup screen, shown in Figure 3.3. Select the type of network (Wireless); then select your network from the Wireless Network list. You'll be prompted to enter your network's password, and then you're ready to rock and roll.

Figure 3.3 *Configuring the TV to connect to your Wi-Fi network.*

To access the Smart Hub, press the Smart Hub button on the TV's remote. This displays a First Screen bar of your most-used apps along the bottom of the screen. You can select an app from here or display the full Smart Hub by pointing to and then clicking the Smart Hub icon within this bar.

The Smart Hub consists of multiple screens for different types of entertainment:

- On TV, which offers suggestions for currently available programming on traditional television. You can use this page to quickly click to view specific programs or to display a more traditional onscreen programming guide.
- Samsung Apps, which is where you access all available web-based content, including streaming video services, social networks, and Skype.
- Games, which provides access to various online games (both free and paid).
- Multimedia, which enables you to access your own digital media stored elsewhere on your home network.
- Movies & TV Shows, which provides suggestions for streaming web-based content.

You'll do most of your browsing via the Samsung Apps screen, shown in Figure 3.4. Here you find apps for all the major streaming services, including Netflix, Hulu Plus, Amazon Instant Video, HBO Go, Vudu, YouTube, Vimeo, Pandora, Spotify, TuneIn Radio, and more. There are also apps for Facebook, Twitter, and Skype (using the TV's built-in camera and microphone). Click to open an app, sign into the service (if necessary), and then start watching or listening or communicating or whatever.

Figure 3.4 *Browsing web-based media from the Samsung Apps screen.*

This screen comes preloaded with some of the more popular apps. You can download additional apps via the Samsung Store, which you also access from this screen.

Operation is via the TV's included remote control, the accompanying smartphone/ tablet app, voice command (the set has a built-in microphone, remember?), or hand gestures. This last one is an interesting application of the set's built-in camera; just point and "grab" to select an item onscreen.

It's all very high tech. The bottom line is that this set, like most other current- generation smart TVs, makes it relatively easy to view just about any type of programming you can think of. It takes a little time and effort to get everything set up properly, but then it's a matter of pointing and clicking to get to what you want to watch.

Exploring Smart TV Set-Top Devices

If you have an older TV (or even a lower-priced newer one without built-in connectivity), you can add similar smart TV features by purchasing a streaming media set-top device. There are lots of these devices, with the most popular being the Roku models, Apple TV, WDTV Live, and Amazon Fire TV. All of these devices are small enough to hold in your hand and sell for $100 or less.

Consider the Roku 2, shown in Figure 3.5. This one's smack dab in the middle of the Roku line (between Roku 1 and Roku 3, naturally), and sells for $69.99. It connects to your home network via Wi-Fi and to your TV via HDMI, and includes its own remote control. Configuration is as easy as navigating through a handful of setup screens.

Figure 3.5   *Roku 2 streaming media player.*

Like all Roku models, the Roku 2 comes with a number of popular apps (they call them "channels") preinstalled, including Netflix, Hulu Plus, Amazon Instant Video, HBO Go, Vudu, YouTube, Vevo, Pandora, Spotify, and TuneIn Radio. You can download a plethora of additional channels online for a variety of different streaming services; because of its popularity, Roku has the most available third-party apps of any of the currently available smart TV devices. (Figure 3.6 shows some of the most popular Roku channels.)

Figure 3.6   *Navigating online content on the Roku 2.*

 Note

If you want to access digital media stored elsewhere on your home network, install the Plex channel. Plex is a streaming media server application you install on the host PC, which then streams your media to the Plex app on your Roku box. (Learn more at www.plex.tv.)

If one of these little boxes is too big for you to deal with, consider a smart TV on a stick. These are streaming media devices in the form factor of a universal serial bus (USB) dongle, such as Google's Chromecast, the Roku Streaming Stick, and Amazon's Fire TV Stick. As you can see in Figure 3.7, these devices plug into any open HDMI connector on your TV and provide similar app functionality for web-based streaming media services. There are fewer cables to worry about, plus the cost is lower, ranging from $35 for the Chromecast to $49.99 for the Roku Streaming Stick. The Roku and Fire sticks come with their own remotes, while you operate the Chromecast with the accompanying smartphone app. It's a nifty way to add smart TV functionality to any TV set that has an HDMI connection.

Figure 3.7   *The Google Chromecast streaming media stick.*

How to Choose a Smart TV or Device

If you're considering the purchase of a new smart TV or device, there are several factors you want to consider—once you get past the basic TV-related stuff, of course.

First, determine whether you really want a new TV or whether a streaming media player connected to your old TV will do the job. You get pretty much the same functionality with a sub-$100 set-top box as you do a $1,000 top-of-the-line smart TV set, so the streaming media player route is a more affordable one. In addition, it's a lot easier to upgrade (re: throw out and buy a new one) a $50 set-top box than it is to replace a $500 or more TV when things change. And things always change.

Whether you're looking at a TV or a set-top box, you want to make sure that the device includes access to those streaming media services you use the most. While virtually all such devices include access to Netflix, Hulu, and YouTube, only a few let you connect to Amazon Instant Video. Most include access to Pandora and Spotify, but less-popular streaming music services aren't always included. Check the available apps or channels to make sure you're happy with the selection.

Next, consider whether or not you want to access media stored on your own home network. If all you do is stream movies and TV shows from the web, this function-ality isn't a big deal. But if you have a large library of digital music, recorded TV shows, or DVD rips, you want to make sure your new smart device can access and play everything you own. Check to see if the device offers streaming over a local network (typically via DNLA), and that it can play back media in the file formats you use. This is particularly important if you have a lot of DVD rips, but can also trip you up with some less-popular digital music format—especially high-resolution formats, such as Flac and Windows Media Audio (WMA) Lossless.

Almost all of these TVs and devices offer Wi-Fi connectivity, which is fine for most households. If you prefer the reliability and speed of a wired connection, however, look for a device that includes Ethernet connectivity.

You should probably look at any additional features offered by a given device. Some smart TVs (but no current set-top boxes) come with built-in cameras and microphones for live video chatting and gesture- or voice-based operation. If this sort of thing is important to you, take it into account.

Now it's time to consider how the thing works—the interface and basic operation. Make sure the device's onscreen menu system makes sense to you, and that you can easily get to where you want to get. Make sure you like how the remote works, or that there's a smartphone app available if you prefer using that. Also, if you have a universal remote for your larger home theater system, make sure it's compatible with the device you're considering.

Finally, there's the price. A streaming stick like Chromecast is the most affordable option, and set-top boxes aren't much more expensive. If you're in the market for a true smart TV, however, be prepared to spend a little more for the "smart" features than you would for a non-connectible model.

And don't forget the cost of the streaming services themselves. You'll pay around $10 a month for Netflix, Hulu Plus, Spotify, and the like. While ten bucks doesn't sound like a lot, it starts adding up when you subscribe to multiple services. Go with a half-dozen services and pretty soon you're spending as much for your online entertainment as you would on a traditional cable bill.

How Secure Are Smart TVs?

Here's a side issue worth considering. Since a smart TV or smart TV device connects to the Internet and has a CPU and an OS, it's just as capable of being hacked as is your typical desktop or notebook computer. You don't think of your smart TV as a computer, but it really is. And just like a computer can be hacked or attacked over the Internet, so can your smart TV.

Hacking Into the System

Why would anybody want to hack your smart TV? For starters, because it stores some interesting personal information, in the form of user names and passwords for all the services you subscribe to, such as Netflix and Hulu. And if you subscribe to Amazon Instant Videos, that's the user name and password for your entire Amazon account. See where that might lead?

And hacking doesn't have to be that malicious. So-called man-in-the-middle attacks place the attacker between the Internet service or broadcaster and the smart TV, enabling the attacker to feed his own content to the victim's screen. Instead of getting the service's normal commercials, then, you may receive commercials from the attacker's company. Not necessarily world endangering, but still not desirable.

In case this seems too theoretical, consider this real-world example of smart TV hacking. In June 2014, Columbia University researchers Yossef Oren and Angelos Keromytis exposed a flaw in the Hybrid Broadcast-Broadband Television Standard (HbbTV) used on millions of European smart TVs. HbbTV has been adopted by 90% of smart TV manufacturers in Europe to add interactive HTML content to terrestrial, cable, and satellite signals. Oren and Keromytis revealed that the HbbTV standard is vulnerable to large-scale exploitations that would be "remarkably difficult to detect."

This so-called "red button" attack, named after the red button on a user's remote control, would enable a hacker to intercept the sound, picture, and accompanying data sent by a broadcast. The attacker then becomes the broadcaster, feeding whatever content he wants to the victim—and receiving data sent by the victim to various smart TV apps. A hacker could use this exploit to display bogus commercials on a victim's TV screen, or log into the victim's Facebook account and post with that person's name.

What this type of attack reveals is the paltry amount of security inherent in this new generation of connected devices. A smart TV (or any smart device) needs to be every bit as secure as your computer system, and most aren't. Where your computer is protected (somewhat) by a firewall application, most smart TVs do not have even this basic level or protection. This leaves them vulnerable to attacks that wouldn't be near as successful on a more secure personal computer.

An Eye Into Your Living Room

Then there are the security issues presented by those smart TVs that include built-in cameras. Imagine a man-in-the-middle attack where an attacker gains control of your TV's camera, and uses it to spy on whatever you're doing in your living room or bedroom. This could be simply voyeuristic or it could let the attacker know when you're out of the house, thus setting you up for potential burglary.

Again, this isn't a theoretical issue. Researchers Aaron Grattafiori and Josh Yavor, security engineers at the firm ISEC Partners, recently discovered a security hole in some Samsung smart TVs (like the one we examined earlier in this chapter) that enabled attackers to hack into the Skype application and remotely turn on and control the TV's built-in camera. That's scary stuff.

Now, to the company's benefit, Samsung promptly sent out updates to its devices to patch this security flaw. And if you're really concerned about your TV spying on you, you can always put duct tape over the built-in camera. Not very elegant, but effective.

Official Snooping

Unsolicited snooping doesn't have to be the province of black hat hackers and the criminal element. It's equally likely that your smart TV's manufacturer is spying on you.

In November 2013, British tech blogger Doctorbeet discovered that his then-new LG smart TV was keeping track of everything he watched. Every time he changed the channel, the activity was logged and transmitted back to the LG mothership. LG then knew every program he watched and could use that data however it saw fit.

LG calls this "service" Smart Ad, because it sells the collected data to advertisers. According to LG, Smart Ad "analyses user's favorite programs, online behavior, search keywords, and other information to offer relevant ads to target audiences." Hoo boy.

Now, there's a setting on that particular LG model called Collection of Watching Info. It's toggled on by default, no surprise, and most people would never get that deep into the menu system to turn it off. Well, Doctorbeet did tiptoe into the menus and deactivated this setting. Unfortunately, it had no effect on the data collection, which continued unabated. So much for viewer choice.

You know what's even worse? This information is sent back to LG in *unencrypted* form. That means any reasonably tech-savvy monkey could intercept the data and know what programs you're watching when. That could be relatively harmless, unless you're watching something that you don't want your spouse or employer or pastor to know about. Scary, eh?

By the way, LG later responded to Doctorbeet's publicizing this issue by changing their terms of service, but not in a good way. Now, if you opt not to agree to this invasion of privacy, LG disables most of the "smart" functionality in your smart TV. That's one way of dealing with the issue, I suppose, but it's not really 21st-century privacy savvy.

Integrating Smart TVs into the Internet of Things

Okay, so it's pretty obvious that the current generation of smart TVs has very little to connect it to the Internet of Things. Just because a TV or set-top device lets you watch both broadcast and Internet-based programming doesn't make it hyper-intelligent or even moderately clever. It just adds more types of programming to what is still more or less a non-participatory device. A TV that can play old episodes of *Doctor Who* on Netflix is still just a TV.

For a smart TV to become truly smart, it needs to do more. Not surprisingly, there are people working on this.

The first thing smart TV manufacturers are likely to do is make it easier to control the smart TVs themselves. Let's face it, picking through the choices on Hulu or searching for your favorite movie on Netflix isn't easily accomplished with a traditional four-arrow remote control. Some manufacturers have experimented with including a full-fledged keyboard in a handheld remote, but that's a little too cumbersome. A better solution might be a touchscreen tablet-like controller, a remote app on a smartphone or iPad, or even Siri-like voice control. Samsung, if you recall, uses its built-in camera to enable rudimentary gesture commands, which is another way to go. Whatever the approach, the smart TV companies need to make it easier to find all the various programming they enable.

Beyond the control issue, future generations of smart TVs are likely to get smarter about what you like to watch. These new smart TVs will collect data about what you watch and when (and, if you have multiple viewers in the same household, which you probably do, what each viewer likes to watch) and make assumptions about your future viewing habits. Even better, your smart TV might connect to your Facebook or Twitter account to discover what shows your friends are watching.

All this data will be assembled and collated, and your smart TV will start making recommendations for future viewing. The set might even go the next step and create a new "just for you" screen with one-click access to the recommended programming, or just set the onboard DVR to record these programs for your viewing convenience. With your smart TV making smart choices about what you want to watch, you'll no longer have to deal with the increasingly Byzantine program guide. You won't have to think about what you want to watch at all; your smart TV will do your thinking for you.

Of course, this type of viewing information can go both ways, so expect your smart TV to feed details of what you watch back to the programming sources—and, more importantly, their advertisers. (As you've read, this is already happening with some manufacturers, such as LG.) This will let them feed more relevant commercials to you and other viewers, so those hip twenty-somethings in the audiences will no longer be subjected to commercials for miracle socks and reverse mortgages. It's all about targeted advertising, based on the data collected by your smart TV.

Future smart TVs may also use their Internet connectivity to overlay related information on the main viewing screen. If you're watching a sporting event, for example, you may see team or player stats superimposed on the screen, or displayed in a side window. If you're viewing a classic movie, you might see bios of the director and stars, with links to other similar movies you might like.

In addition, expect smart TVs to include more interactive chat capabilities. When you're watching a movie or show, you'll be able to tweet or post to Facebook about what you're watching, and participate in group chats about the show. These might be video chats, conducted in a pop-up window and enabled by your set's built-in camera.

Future iterations of smart TV will turn the TV set into a hub for a variety of household activities. For example, you might feed video from your home's security cameras to your smart TV, so you can see who is ringing your doorbell or if your baby is asleep in her crib. (To be fair, this capability exists today in a lot of high-end, whole-house audio/video/security systems but is sure to trickle down to more affordable systems in the future.)

You can use that big TV screen to view all sorts of information. Why not click an onscreen button to view a graph of your home's water or energy usage? Or display a map that shows where all the members of your family are at the moment? Or a diagram that shows which rooms are occupied and where the lights are still on? Or a live feed from inside your refrigerator that lets you know if you have a cold beer waiting for you?

Imagine using your living room TV to control various household operations. Just point and click at the screen to turn on the lights in a given room, start the oven or dishwasher, even enable your outdoor sprinkler system. You're in front of that screen a lot of hours during the day; why not use it as an interactive household controller?

There's no reason why your smart TV can't be the main controller for all your household operations. It's right there in front of your couch, where you're no doubt sacked out. There's no reason to get up to turn up the heat or turn down the lights; you can do it all from the controller interface built into your smart TV. And your TV can alert you when things are amiss anywhere in your house or on your property.

The security angle is key. Not only can you use your smart TV to view real-time data collected from your home security system and live feeds from various security cameras, it can also interact with other devices to provide more intelligent analyses. Imagine a system that uses face recognition to learn what each member of your family looks like; the system could then look at the faces in the security cameras and alert you when a stranger is at the door or in the house. (And, conversely, not bother you if it's a known person raiding the fridge.)

In short, there's a lot more that your smart TV can do than what it's capable of doing today. Just wait for it.

SMART TVS AND YOU

Now that you know all about the past, present, and future of smart TVs, it's time to decide whether a smart TV or smart TV device fits in your current lifestyle. Should you buy a smart TV today—or wait for some future iteration?

Let's face it, today's smart TVs are just a way to obtain more and varied programming than you get from your cable or satellite company. Whether you want to completely cut the cable cord or supplement your 400+ cable channels with a similarly large assortment of Internet-streaming sources, a smart TV lets you do it.

The primary selling point of today's smart TVs is that they integrate programming across multiple sources. No longer are you limited to just broadcast or cable programming; with a smart TV, you can easily switch from watching your daily fix of Jimmy Fallon on *The Tonight Show* on network television to binge-watching the latest season of *Orange Is the New Black* on Netflix.

If you want the additional programming that's available online, a smart TV is a better way to go than watching the same programs on your notebook or desktop computer screen. Yes, you can view Netflix and Hulu on your handy dandy PC or tablet or smartphone, but movies and TV shows are made to be watched on big screens. Computer- or tablet-based viewing, while fine for college students in their dorm rooms, doesn't cut it for a modern family accustomed to widescreen entertainment.

So if you're a Netflix or Amazon or YouTube junkie, you need some sort of smart TV device. That doesn't have to be a literal smart TV, of course; it could also be a smart TV device in the form of a set-top box or streaming media stick. These add-on devices are a lot lower-priced than even a 32" smart TV and can easily be replaced when they break or become outmoded. As long as you have an open HDMI connector on your TV, it's easy to connect one of these little devils.

That said, a full-fledged smart TV can offer more functionality than what you get in a set-top box or dongle. A built-in camera lets you use your smart TV for Skype and other video chat, and that's kind of cool to do from your living room couch. In addition, easy as the add-on devices are to use, it's just a little simpler to do everything from the TV itself. Some people value that ease of use.

Of course, if you're happy with broadcast or cable television and don't care about Netflix and those other services, a smart TV doesn't make a lot of sense. Save your bucks and stick to a conventional TV, at least for the nonce.

Flash forward three or five years, however, and those future smart TVs will offer a lot more functionality than just an easy way to watch shows on Netflix. Truly smart viewing recommendations, interfacing with and controlling other household devices and so forth, offer the type of promise inherent with the Internet of Things. When your smart TV becomes more than just a passive viewing device, things get really interesting—and worth your active consideration.

4

Smart Appliances: From Remote Control Ovens to Talking Refrigerators

Smart appliances represent a large potential part of the Internet of Things. In the future, smart refrigerators might monitor food usage, and text or email you detailed shopping lists; smart ovens will know what you're cooking and when you eat, and turn themselves on automatically; and smart laundry machines will run when you're away from home and won't hear the noise. But there's a lot more to smart appliances than just Internet-connected dishwashers and intelligent toasters, as you'll soon see.

Understanding Smart Appliances Today

There's nothing quite so dumb as those big metal boxes taking up space in your kitchen. Your refrigerator is just a big insulated box designed to keep food cool. Your dishwasher has only enough brains to start the designated cycle at the time you set beforehand. Your oven has a timer, too, but you have to manually set the desired temperature. Your microwave oven is a little smarter, in that it has a few preprogrammed power levels for specific types of food, but the clock doesn't know enough to reset itself for Daylight Saving Time. And while your washer and dryer include some preprogrammed cycles for different types of fabrics, and the dryer maybe even has a moisture sensor to tell you when everything's properly dry, both are still simple machines, waiting for you to tell them what to do.

And that's the state of smart appliances as of a year or two ago: lots of appliances, not a lot of smarts.

As you've probably suspected by now, the use of the term "smart" is as much a marketing ploy as it is a true descriptor for anything connected to the Internet of Things. Different companies define smart in different ways, as do different industries. A smart TV might be more or less intelligent than a smart water softener or smart car, but they're all equally smart in the eyes of the marketing guys.

So what does it mean when we call an appliance "smart"? It all has to do with automating routine operations.

Smart Operation

The first thing that many will find smart about smart appliances is the ability to manage them remotely. We're used to setting a timer or a given start time to fire up the oven or activate the dishwasher; this sort of timed start is useful for starting dinner before you get home from work, or running water-intensive appliances later at night when nobody's taking a shower.

With smart appliances, this automatic operation gets more flexible. Instead of manually setting a timer on the front panel of a device, you can use an app on your smartphone or tablet to remotely press the start button on a given appliance. Some apps even let you program operations in advance—like setting a manual timer, but more sophisticated.

For example, General Electric's (GE's) Brillion smartphone app, shown in Figure 4.1, offers remote operation of select wall ovens. You can be miles away at work, with a big pan filled with a roast in the oven (you put it there before you left in the morning), and all you have to do is tap the app to start up the oven and start cooking. Or maybe you want to preheat your oven to a certain temperature for use when you get home. In any case, it's easy-as-pie remote operation, all thanks to a convenient mobile app and Internet connectivity.

Figure 4.1   *Control your GE wall oven with the Brillion app.*

Even better is LG's new HomeChat app, which lets you send text messages to your appliances just as if they were human. (HomeChat operates via the popular LINE messaging app.) As you can see in Figure 4.2, you use text messages to send commands to your appliances in plain English—and receive messages from them via text, as well. Right now, you can text basic operation commands, but you may eventually be able to text your fridge, "Is the milk fresh?" and receive an answer in return.

Smart Monitoring

For a smart appliance to be truly smart, it must monitor its environment and operation to let you know about things you need to know about. That means sending out some sort of alert or notification when a given operation is done or when something unexpected happens.

Today, appliances typically alert you with a loud buzzer or rinky-dink snippet of music. (I absolutely hate the notification music from our LG laundry equipment!) Smarter appliances will send out smarter notifications.

Figure 4.2   *Chat with your appliances with the LG HomeChat app.*

For example, a smart washer might send a notification to its corresponding smartphone app when the washing cycle is complete. Or maybe you'll get a text message notifying you that your dinner is finished cooking in your smart oven. Or how about an email alert if somebody leaves the refrigerator door open?

The key is to use connected technology to notify you of important stuff happening in the kitchen or laundry room. We have the technology; we can do this.

For example, Whirlpool's My Smart Appliances app, shown in Figure 4.3, lets you monitor the status of all your appliances on your smartphone screen. You can see how much time is remaining for your loads, get notified when a washing or drying cycle is complete, or see how cool your refrigerator is today. It's a lot better than hanging around the laundry room waiting for the buzzer to sound.

Smart Energy Savings

Some of today's smart appliances use their smarts to cut down on energy usage. If an appliance knows when power consumption is lower or rates are cheaper, it can program itself to operate during those times. In addition, smart appliances typically include other energy-efficient functionality, such as a dishwasher using less water per cycle or a refrigerator incorporating more insulation to keep things frosty.

Figure 4.3 *Monitor your laundry equipment with Whirlpool's My Smart Appliances app.*

For example, LG's Smart Grid technology detects when local power consumption is lowest and schedules more operations during that time. Figure 4.4 shows one of the Smart Grid configuration screens. (Note that this feature is only available to those homes served by a Smart Grid power company.)

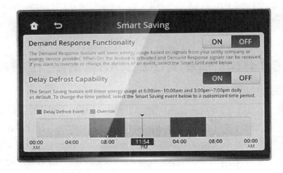

Figure 4.4 *Increase energy efficiency with LG's Smart Grid technology.*

Note

A Smart Grid is a new type of electrical power grid that incorporates built-in intelligence and digital communications technology. Learn more in Chapter 13, "Smart Cities: Everyone's Connected."

Smart Maintenance

Then there's the process of keeping your smart appliance up to date and in tip-top operating condition. This involves having the right sensors within each appliance to determine when some maintenance needs to be done or when some function isn't properly functioning. Then, instead of just flashing a light on the appliance's front panel, you get notified (via app, text message, or email) about the issue at hand. Ideally, the message includes advice or instructions for what you should do next.

For example, you're probably used to seeing an alert light above your refrigerator's water dispenser when the water filter needs to be changed. With a smarter refrigerator, you'd get a text message or email to that effect instead. Same thing when you need to add more fabric conditioner to your smart dryer, or if the hot water in your dishwasher isn't getting hot enough.

It's all about smart diagnostics and alerts making you more aware of things you need to be aware of—and aiding in the diagnosis when things go wrong. For example, LG's Smart Diagnosis app, shown in Figure 4.5, helps you troubleshoot issues by transmitting relevant data over Wi-Fi to the related smartphone app. This data can then be passed on to LG's Call Center for their technicians to analyze. It certainly reduces the guesswork.

Smarter Food Storage with Smart Refrigerators

Initial smart refrigerators were simply fridges with liquid crystal display (LCD) screens in the door. That's nice, especially if the screen (and the fridge) are connected to the Internet, so you can do your web browsing (for recipes, let's say) or watch Netflix while you're cooking. But the extra cost associated with this feature made these models price-prohibitive; you could get the same functionality with a $200 Android tablet.

That's starting to change and will change even more in the months and years to come. We're talking about true added functionality—especially as your smart refrigerator starts to connect to other appliances in the kitchen.

Figure 4.5   *Diagnose problems with LG's Smart Diagnosis app.*

First, the basic stuff. A smart refrigerator should be able to monitor its own performance and keep itself at the proper temperature for the food inside. If the temperature changes, it should notify you before your food goes bad.

It's also nice if the smart fridge "nudges" you to do certain things in certain situations. How about a text message or alert from the smart appliance app when someone leaves the refrigerator door open? Or when the filter for the in-door water dispenser needs to be replaced?

Beyond that, how about a refrigerator that lets you know when foods expire? Or one that knows what foods you like and monitors its own inventory? Even better, the smart fridge should know when you're low on your favorite foods and beverages, and use that information to assemble a grocery list. Even better than that, the fridge should then send (via text or email or whatever) the grocery list to your local grocery store, which then fills the order and delivers those groceries to the door—and sends an electronic invoice for the goods purchased. (You'll still have to manually stock the groceries in the fridge, however, unless someone invents a handy dandy robot to do that chore.)

This communication doesn't have to be all high tech, either. If you're there in the kitchen, why not have your refrigerator tell you (in one of those politely clipped British accents, probably) that you forgot to close the door, or that it's time to preheat the oven, or that you're out of mustard? Maybe all this smart stuff will be more palatable if it comes with the right accent.

Connect your smart TV to other devices in your home and things get even more interesting. Imagine that you're watching your favorite cooking show on the Food Network and see a recipe you like. You use your smart TV to save the recipe and send it to your smart refrigerator. The fridge stores the recipe and determines whether or not you have all the ingredients necessary. If you don't, it either tells you what you need or automatically adds those items to the next grocery list. And all this happens before the show you're watching goes to commercial.

Even better, why not let your smart fridge figure out what to cook, based on what ingredients you have on hand? Using a combination of sensors and internal cameras, your refrigerator will know what's inside and use its own built-in intelligence (matched to a database of recipes, of course) to figure what you can cook, based on how well stocked it is. Takes all the guesswork out of meal planning.

Within the kitchen, connect your smart refrigerator to your smart range to make cooking easier. Send a recipe from your smart refrigerator to your smart range to automatically set the precise cooking time and temperature, and then get alerted when it's time to start cooking.

What about using a smart fridge to help you diet? Connect your refrigerator to your smart scale or fitness band and it'll know what foods you should be avoiding. Imagine your smart refrigerator nagging you not to eat that pint of ice cream or slice of pie as you start to take it out of the fridge. Annoying, perhaps, but also nudging you toward a healthier lifestyle. (Or maybe the fridge is so Draconian it refuses to stock unhealthy items? It's possible.)

If your smart refrigerator has a built-in LCD touchscreen (and it probably does), then you can use that touchscreen for all manner of applications. You can view the daily weather report, read the latest news, even watch your favorite TV morning show. You can also use the screen to display a slideshow of family pictures or the built-in speakers to listen to music via Pandora or Spotify. And, of course, surf the web for the latest recipes and cooking tips.

Once you start thinking outside the (ice) box, the sky's the limit. Ashley Legg of Yanko Design envisions a Smart Fridge where the entire refrigerator door, when closed, becomes a touchscreen interface, thanks to electrochromic window technology. Essentially, you have a glass door that becomes opaque at the touch of a button; combined with touch sensor technology, you get a large operating interface, as shown in Figure 4.6. You can then use the touch interface to display menus, ingredient lists, step-by-step instructions—you name it.

But that's the future. Today's smart refrigerators incorporate self-diagnostics, "smart" cooling (setting different temperatures for different areas of the fridge), and some type of touchscreen display. The bigger displays function like embedded tablets, with full Internet connectivity. But there's not a lot of smart connectivity as yet; that functionality is still to come.

Figure 4.6 *Yanko Design's Smart Fridge concept.*

 Note

Because of the combination of high price and limited functionality, smart refrigerators (sometimes called Internet refrigerators) have not seen much success in the marketplace. Whirlpool recently discontinued their lone smart refrigerator model, and LG dropped their entire line of ThinQ smart appliances. I expect all these companies to keep plugging at the smart appliance market, but it's going to take more than a fancy touchscreen to pique consumers' attention.

Smarter Cooking with Smart Ovens

There's a world of possibilities when you add just a little intelligence to your oven and cooktop. It's all a matter of trying to make cooking easier—and more accurate.

First, there's the simple matter of monitoring cooking temperature. Put a temperature probe or temperature/humidity sensors in the smart oven, and you can be alerted (via smartphone app) when your casserole or turkey reaches the correct temperature. Or with a little predictive intelligence, you get an alert when the meal is 10 minutes away from being done, so you'll have time to set the table.

Of course, if you have that range app on your smartphone or tablet, you can use it to start up your oven or cooktop and dial in the correct cooking temperature. This will let you warm up your oven or even start cooking while you're still at work or sitting in rush hour traffic on the way home. (Figure 4.7 shows a GE double convection wall oven with built-in Brillion smart operation.)

Figure 4.7   *GE's Brillion-enabled smart wall oven.*

Looking beyond that, there's a lot more you can do if you connect your range to your refrigerator or even to a smartphone recipe app. The most obvious application is to select a recipe on your phone or on the smart fridge, and then have your oven automatically set itself for the correct cooking temperature and time.

And then there's Whirlpool's touchscreen cooktop prototype, shown in Figure 4.8. This one's kind of cool, as it melds a touchscreen display with a glass induction cooktop. The surface stays cool to the touch, as the induction heating elements only interact with metal pots and pans. Display your recipes or other info on the cooktop itself and control things with a swipe of your fingers.

Smarter Cleaning with Smart Washers and Dryers

How smart do a washer and dryer need to be? All they do is wash and dry clothes, right? Well, there's a lot you can automate about your laundry operation—and become more efficient, to boot.

Figure 4.8 *Whirlpool's touchscreen cooktop prototype—cool to the touch.*

First off, the smarter the equipment is, the better it can wash and dry your clothes. Moisture sensors in both washers and dryers can help fine tune wash and dry cycles based on the clothing being laundered. Automatic sensing of fabric type can help the washer set the correct wash cycle and the dryer adjust the heat accordingly.

Of course, there's always remote control operation. Use your smartphone app to start your washer remotely, even when you're not at home. You can also use the smartphone app to monitor your laundry's progress and get notified when the load is done.

And don't forget energy usage. A smart washer will know when's the best time to operate based on water usage and rates, and the dryer will know the same based on electric rates. Smarter operation will reduce your energy costs.

Connect your washer and dryer together and the process gets even more efficient. For example, select GE laundry equipment, such as the washer in Figure 4.9, uses CleanSpeak technology to pass information from washer to dryer concerning how much moisture is left in the clothes. The dryer can then adjust the drying cycle accordingly.

Figure 4.9   *This GE washer uses CleanSpeak technology to communicate with the matching dryer.*

Whirlpool takes this smart business a step further with its Smart Front Load Washer and Dryer that don't connect to each other, but rather to the Nest Learning Thermostat. The Nest thermostat is a smart thermostat that knows when homeowners are away and feeds that information to the Whirlpool laundry pair. The washer and dryer then switch to lower-power operating modes to save energy. Expect more smart appliances to interface with smart thermostats and other smart devices, especially when energy savings can result.

Smarter Dishwashing with Smart Dishwashers

Most of the functionality possible in smart washers is also possible in smart dish-washers. You can operate the dishwasher remotely via a smartphone app and be notified when a load is complete. In addition, you can remotely lock the dish-washer, to ensure that your kids won't be playing in it when you're not there. (The Whirlpool dishwasher shown in Figure 4.10 features this type of remote operation.)

In addition, the smart dishwasher will know when the best time is to operate, based not only on water and electrical usage and rates (when connected to a Smart Grid), but also on when the house needs to be quiet, when the tub is filling up, and when you need those dishes for the next meal. Ideally, it will also determine what type of cycle to use, by sensing the types of dishes and pots and pans inside, and how dirty they are.

Figure 4.10   *Operate this Whirlpool dishwasher via smartphone app.*

SMART APPLIANCES AND YOU

As you've seen, most of today's big appliance manufacturers—including Whirlpool, GE, LG, and Samsung—include smart functionality on many of their high-end models. You won't find these features on the bargain-priced models you see advertised in the Sunday paper, but if you step up a few hundred bucks, things start to get a little smarter.

Despite these companies' best efforts, smart appliances have been a tough sell. They've been trying to integrate smart technology into their appliances for several years now, but consumers just aren't buying. In 2012, smart appliance sales totaled just $613 million worldwide, according to Pike Research. That's a drop in the bucket to the $184 billion worth of appliances sold that year. (Although Pike projects a hockey-stick increase to $34.9 billion by 2020—which may or may not be reasonable.)

Why is it that you likely don't have any smart appliances in your kitchen? There are a number of valid reasons.

First, and perhaps most important, smart appliances are expensive. Very expensive. If you're in the market for a new refrigerator, you might have a budget of $1,000 or so for a reasonably higher-end model. If you want a smart fridge, however, be prepared to shell out $3,500 or more. That's an incredibly large price differential, and more than most consumers are willing to consider.

Second, the appliance replacement cycle is much longer than for typical consumer electronics items. You might replace your smartphone every two years, your PC every four or five years, and your TV every eight to ten. But your refrigerator or laundry pair? Think fifteen years or more in between purchases.

And you don't replace your appliances because they're losing their luster. You replace them because they wear out. A twenty-year-old refrigerator cools your food just as well as a shiny new model. There's little to no technological reason to upgrade your appliances on a more rapid cycle, so you don't.

This brings us to the question—should you invest in new smart appliances? In and of themselves, probably not; that is, you shouldn't (and probably won't) buy a new fridge or clothes washer unless your old one breaks down. Appliances are major purchases, and they're built to last. You don't buy new ones on a whim.

That said, if you do need a new dishwasher or range, you should probably look at those smart models available today. In some instances, a smart appliance may be worth the extra cost—especially if you're already considering a high-end model. But for most of us, the smart features just aren't smart enough yet to warrant the additional expense.

This may and probably will change in the future, as costs drop and functionality increases. It's almost inevitable that the kitchen of the future will be connected in a way we can only dream of today. We just have to get there.

So unless you're forced into a new appliance purchase in the near future, taking a wait-and-see attitude may be best. Smart appliance technology is progressing rapidly, and the options available five years from now may be much more appealing.

If you're in the market for a new appliance before then, consider these smart features now or soon to be available:

- **Smart cycles**—For washers, dryers, and dishwashers, look for models with dozens or even hundreds of customizable washing/drying cycles, dependent on different types of loads. Some models will automatically customize cycles, others will offer additional cycles for download from their websites. However, if all you use is the Normal cycle, this may be overkill.

- **Smart food management**—We're talking refrigerators here, of course. If your fridge can keep track of its contents, send alerts about expiration dates, and let you know when supplies run low, that's a good thing. Don't expect today's current generation of smart refrigerators to do your ordering for you, however; that's several years out.

- **Smart diagnostics**—It's nice if something goes wrong for your big-ticket appliance to tell you (or your repair person) exactly what's wrong. It certainly makes fixing it easier.

- **Smart grid usage**—If your power company offers Smart Grid technology, you can take advantage of it with a Smart Grid-compatible appliance that can help you save a few dollars on your utility bill.

- **Smart operation**—From front-panel touchscreens to remote-control smartphone apps, look for ways to control your appliances that match your lifestyle. Make sure it's something you'll actually use, however, and not just something that looks cool. (Although looking cool is good, too.)

5

Smart Homes: Tomorrowland Today

Smart TVs and appliances are just part of the smart home of the future. Already you can control heating, cooling, lights, and alarm systems over the Internet with your smartphone; imagine a world where your home is smarter than you are and can do all of this (and more!) automatically. It's what Nest (a company we discuss later in this chapter) calls the "conscious home," and it's coming your way soon.

Automating the Home

When you assemble a collection of smart devices under one roof, and you enable them to connect to and communicate with one another, what you end up with is typically called the *smart home*. Some people also call it *home automation*, because all of your smart devices work in concert to automate a variety of household chores and operations.

Home automation has been around for a few decades now, in terms of automating lighting, heating and cooling, and the like, typically in high-end full-house systems created by suppliers such as Crestron and Control4—or in lower-end systems, such as the X10.

The concept of home automation is a simple one. You take a given task or operation, typically performed when you manually turn on a device, and somehow make that device turn on and work automatically, without your manual intervention. Simple home automation uses timers and clocks to enable the operations; more advanced home and/or smart home technology can handle more sophistication operating scenarios and trigger devices based on input from other devices.

What kinds of smart things can you find in a smart home? Just about anything that plugs into an electrical outlet is game, which means centralized control of lighting, heating and air conditioning (A/C), appliances, door locks, motorized blinds and curtains, home security systems, and the like.

The primary benefit of smart home technology is that of convenience; by automating basic operations, you don't have to be bothered with them. There are other benefits, of course: faster response to changes in the environment or breaches in security, more efficient operation leading to energy savings, and the like. But let's be honest; the real reason you want to automate your home is because it's really, really cool to control things from your smartphone or have them turn on and off automatically based on various inputs. It's kind of like having the Clapper, but on a more advanced level.

Not surprisingly, there has been an increased interest in home automation of late, as more and more smart devices become available at more affordable costs. Thank the Internet of Things for enabling smart home technology for everyone.

Convenience

A smart home is more convenient than a regular home. There's less you have to do and to remember to do; you let the home (or rather, the smart devices in the home) do most of the work for you.

Even a relatively dumb home has some smart features. Consider your basic thermostat, which you program to turn your air conditioner and furnace on and off at specific times, or to cool or heat your home to a given temperature. Or your coffeemaker, which is programmed to turn itself on before you wake up so you'll have a fresh cup of joe waiting for you. Or maybe you have a home security system programmed to dial the authorities if someone breaks in. Basic stuff, yes, but still automation.

Now take all these basic operations and dial them up to a higher level. Instead of just controlling your furnace and air conditioner, your smart thermostat communicates with your garage door opener to know when you've left home, and then with your washing machine to know that it's okay to start washing but to use the "Away" cycle. Instead of just brewing a cup of coffee when you wake up, your various connected devices gently turn up the lighting, ramp up the furnace, and start the shower. Instead of just sounding the alarm when an intruder is detected, your home security system locks any open doors, broadcasts a diagram of your house to the police department, and locks down anything nearby.

Then there are those things that weren't automated before but now are part of the smart house operations. Light-emitting diode (LED) lighting is automatically dialed up and down depending on who's doing what and where; if you're watching TV, the automatic lighting is different from when you're reading a book or talking on the telephone to friends. If the house is too cool, use your smartphone to crank up the heat. You can also use your smartphone to open and close your motorized window coverings, lock the doors after you've left for the day, and check in on your kids as they sleep.

 Note

Smart technology enables you to create "scenes" for various activities. It's more than just setting ambient lighting; press a button for a specific activity and the lights, temperature, and background music are adjusted to set the scene.

The ability to control everything in your home from your smartphone or computer—or to have everything triggered automatically, based on parameters you set in advance: It's an appealing prospect.

And that's before the robots get involved. Okay, they're already involved, to a small degree, in the form of carpet- and floor-cleaning robot vacuum cleaners, such as the Roomba. In the smart home of the future, these relatively dim-witted robotic cleaners will be supplemented by robots that bring you snacks and drinks, clear dishes off the table, and even transport you to another room when you want. Jane, stop this crazy thing!

Security

Just about any home security system more advanced than a barking dog is, to some degree, automated. Your current system probably incorporates window and door sensors, and maybe some motion detectors. When a sensor detects something abnormal, the system phones the system office which then determines the extent of the problem and calls the police if necessary.

Smart home security systems add a lot more options to this type of basic system, such as webcams throughout the house that you can monitor from your smartphone or computer. The system also coordinates with other home systems so that it knows when you're away (thank you, smart garage door opener or smart thermostat) and activates the system automatically. Of course, you can also activate or deactivate the system via smartphone and monitor the status via your mobile app as well.

The smart system also ties into your home lighting, both inside and outside. When you drive out of the garage, lights start to turn on and off throughout the house, simulating normal at-home activities. Front path, garage, and kitchen lights turn on when you arrive home at night. Backyard lighting activates if unexpected activity is detected. And you get a message on your smartphone app if anything unusual happens.

Smart home security goes beyond burglar alarms, however. A whole house system incorporates smart smoke and carbon monoxide detectors directly wired into your local fire department; sensors on your water heater, water softener, and sump pump to alert you of liquid incidents; and a one-button emergency setting that sets off the alarms, turns on all the lights, and dials the police department when you feel threatened.

In addition to this kind of whole-house security, there's the security of knowing that everything in your house is working correctly. Your furnace can send an alert message to your smartphone when the filter needs replacing, or your refrigerator can text you when it needs service. If something goes wrong with a critical device when you're away from home, it can contact the appropriate service center to effect repairs without you even getting involved.

 Note

Smart home technology is particularly useful for the elderly and disabled. By automating a variety of functions—and providing enhanced monitoring—smart devices can make it possible for those who might otherwise require caregivers or institutional care to continue living in their own homes.

Efficiency

Your smart home knows more than you do about saving energy. You're the guy who leaves all the lights on when you leave a room; your smart home is smarter than that. Automatic lighting turns on and off not only in response to the amount of light outside, but also to whether there's anyone actually in the room or not. Smart appliances are scheduled to run when energy rates are their lowest, and not at all when water or electricity are scarce. Some lights are even programmed to operate at a slightly lower wattage than normal, so that you save energy without even knowing it.

Tying It All Together

For all these smart devices to create a smart home, they have to be connected together. In the old days, home automation meant creating some sort of wired network, often via your home's existing power lines. Today, however, wireless is the way to go; it's a lot easier to set up a wireless network than a wired one.

You might think that smart home devices would just connect to your home Wi-Fi network, and some do. But most smart home companies have invested in their own proprietary networking technologies—which, being proprietary, only work with devices from a given manufacturer. This discourages mix and matching among smart devices, but also complicates your life to some degree. Once you settle upon the smart hub or technology you like, you're pretty much locked in from there.

A Short History of Smart Homes

The concept of the smart home isn't a new one, as the automation of household chores has been a common theme throughout science fiction writing and cinema for at least a century. Pulp fiction in the early years of the 20th century was rife with domestic robots and self-servicing kitchens. Ray Bradbury's 1950 short story, "There Will Come Soft Rains," depicts the many labor-saving devices in what we'd today call a smart home, as does his other 1950 work, "The Veldt" (with its subtly malevolent "Happylife Home"). Fast forward to the 1960s and you get the ultimate in home automation with *The Jetsons* cartoon and their robotic maid Rosie. The 1977 thriller *Demon Seed* was all about a smart house that went a little off the rails. And today's *Futurama* TV show envisions all sorts of wacky home automation devices.

But that's all fiction. The reality of automating household chores tracks more closely with the widespread introduction of electricity and electrical home appliances between 1915 and 1920. Before then, well-heeled households automated their chores by hiring domestic servants. As the employment of servants dropped in the new century, households instead purchased new-fangled electrical appliances, such as blenders, mixers, toasters, and the like.

Fast forward a few years to the 1933-1934 Chicago World's Fair which featured the "House of Tomorrow," one of the very first prototypes of what we today would recognize as a smart home. The House of Tomorrow, in addition to being a marvel of modern design (12 sides and 3 stories in a design reminiscent of a towered wedding cake), included a built-in dishwasher, electric lights with dimmer switches, electric garage door opener, and central air conditioning, all futuristic for the time. It also featured passive solar heating and garage space for both the family automobile and family airplane.

By the time the 1962 Seattle World's Fair rolled around, the "modern living" of the future was envisioned as a relatively affluent lifestyle enabled by home automation. We're talking houses with two swimming pools, both requiring little maintenance because they're "automatically vacuumed." Inside, the home has become a giant computer that can "program meals, balance the checking account, and call the library and the grocery store." All clothing, plates, cups, and cutlery are disposable, of course. Plus, food is taken in pill form.

While that level of "modern living" didn't quite materialize (no roast beef and mashed potato pills, sorry), the idea of the automated home continued to flourish—and become a reality. In 1975, the Scottish company Pico Electronics began work on what they called the X10 project. (Nothing special about the name; it was simply the tenth project the company had worked on.) X10 was—and still is—a home automation technology that uses existing home electrical wiring to transmit signals to control lights and appliances and such.

In the X10 system, all appliances and electrical devices are receivers, and the items you use to control the system (remote controls, keypads, and the like) are transmitters. You install the necessary receivers and transmitters to automate a given series of tasks.

The first X10 products reached the market in 1978, sold in mainstream stores such as Radio Shack and Sears. For the first time, tech-savvy consumers could automatically control items in their homes. It was a big deal.

Home automation continued to advance, although the phrase "smart house" wasn't coined until 1984 (by the American Association of Homebuilders). The cost of electronic home controls fell throughout the 1980s and 1990s, thanks to the invention of the microcontroller. Remote control and intelligent control technologies began to be adopted by the building industry and by some appliance manufacturers. Automation technology also found its way into high-end home theater and home automation systems.

Despite all this interest, however, home automation or smart home technology remained—and still remains—primarily limited to use in higher-end systems and devices. The technology is still considerably high-priced, and the lack of a single simple protocol makes the whole thing confusing to consumers. That's starting to change, of course, but there's still considerable room for more affordable pricing and improved ease of use.

Smart Steps to a Smart Home

We won't get to a full-fledged smart home in one fell swoop. Instead, as with most technology, we will take baby steps toward the home of the future, enabling new functionality with the adoption of each new level of technology.

Experts differ as to the levels of smartification that we'll experience, but I see a cool half dozen. They range from basic communications to taking informed action.

Step 1: Basic Communications

The first step in creating a smart home is enabling members of the household to communicate with other people outside the home. Basic to achieving this step is some sort of communications technology, ranging from landline telephone service to mobile phone service to broadcast television reception to an Internet connection.

Obviously, Internet connectivity enables data communication in addition to voice communication, so it's particularly important in terms of the tasks ahead.

Step 2: Simple Commands

We're not talking voice commands here, just the ability to issue some sort of command to perform basic tasks, such as locking or unlocking a door, turning lights on or off, checking for mail, even summoning help when someone falls and can't get up.

At this level, the house will also respond to commands from outside the home. When postal mail is delivered to the outside mailbox, a light goes on inside the house. When someone walks by an outdoor motion sensor, an alarm sounds. When the alarm company is notified of a monitored door being opened, it calls to see if you did it or if there's an intruder in the house.

Step 3: Automating Basic Functions

At this level, manual control gives way to automatic control, via programmed instructions. We're talking about automating such functions as controlling room temperature, cycling lights on or off at specified times, activating or deactivating the alarm system on a given schedule, running the outside sprinkler system according to a set program, and the like.

It's a matter of using technology, typically in the form of programmable timers, to do what you'd otherwise have to do manually. Instead of manually switching on this light or pressing that start button, the task starts whether you're there or not. You still have to tell the device to do its thing, but you tell it well in advance, based on the schedule you enter.

Step 4: Tracking and Taking Action

Now we get to the stage where your home starts watching you. Thanks to development in sensor technology, your home can now track what you do (and where) to determine activity patterns, sleep patterns, even your health status. Your home is converted into a giant monitoring system, and you're the subject.

What does your home do with all this data it collects about you? It uses programmed algorithms or basic artificial intelligence to make decisions based on that data. If your house knows you always get up at 6:00 a.m. during the week, it knows to turn up the heat, turn on the coffeemaker, and so forth. If you exhibit behavior contrary to your norm, your home knows to alert authorities that you may be sick or injured or whatever.

And it's not just taking action based on your activities. At this level, your home can act on unusual data from sensors throughout the house. If the temperature spikes in one room that might indicate a fire, and the house will sound the alarm, turn on the sprinklers, and call the fire department. If the water pressure suddenly drops, the house knows that there's probably a water leak, and it turns off the main line and calls a plumber.

Step 5: Prompting Activities and Answering Questions

At this level, your smart home knows more about you than you do. It will prompt you to take your daily medication, ride your exercise bike, feed the dog, and set out food for dinner preparation. You'll even be prompted to make any necessary phone calls, write a note to your son's teacher, and the like.

Given the inherent intelligence (and access to not only your household information but also the wide world of information available on the Internet), you'll be able to ask basic questions and get accurate answers. Ask your home, "What's the weather going to be like today?," "Do we have all the ingredients for the dinner dish I'm preparing?," "How many days till my wife's birthday?," and "What appointments do I have today?"

Step 6: Automating Tasks

The more your home knows about you and your activities, the more it can do for you; you won't even have to ask the questions. Your home will automatically schedule necessary maintenance and repairs for all its component pieces and parts, reorder medications before you run out, prepare grocery lists (and send them to the grocery to fill and deliver), run the robotic vacuum cleaner, and do just about anything else that it can do autonomously.

And this is the nirvana of smart home technology, at least as we envision it today: a house that is appropriately wired and necessarily informed, making its own decisions, and performing all manner of menial and important tasks. When your home becomes this smart, you can just sit back and enjoy it—or feel free to take a vacation, knowing that your home will run just fine without you.

Simple Components for a Smart Home

What technologies are required to create the smart home of our dreams? There are a number of items that are necessary—most of which we have in our possession today.

Sensors

As with all things in the Internet of Things, we need sensors to determine the status of our environment and of various devices. These sensors can be freestanding or built into other devices.

In a smart home, sensors are needed to detect temperature, humidity, light, noise, and motion. Specialized sensors detect smoke and carbon monoxide levels; proximity sensors detect whether doors and windows are open or shut. Sensors can also detect the status of a given device (on, off, and so on) and the location of devices and people. (And pets, too.)

Note

There are many different types of sensors that may find their way into your smart home. Companies such as Adafruit (www.adafruit.com) and SparkFun (www.sparkfun.com) are big in the sensor business, offering modular-sensing components such as microphones, cameras, thermometers, humidity and barometric pressure sensors, fingerprint scanners, motion sensors, luminosity (light) sensors, force resistors, infrared proximity scanners, magnetometers, accelerometers, and more.

Controllers

Controllers are necessary to send signals to other devices to initiate some sort of operation. In a smart home, a controller can be dedicated to a specific operation, or be part of a larger device (such as a smartphone or home computer) to control multiple devices.

Actuators

An actuator is typically a mechanical or electrical device that actuates a given activity. We're talking motors and switches, such as those in electronic light switches, motorized valves, and the like. Without actuators, nothing gets done.

Buses

A bus is the communication system that transfers data between devices in a smart home. Different types of buses transfer data according to different protocols; for devices within the home to communicate with one another, they must all be compatible with the same bus type.

Interfaces

An interface enables the communication between different devices or between humans and devices. In device-to-device communication, the interface is really just a digital protocol. In the case of human-to-device communication, the interface typically includes some sort of controller and display, so that the person doing the controlling can see what he's doing.

Networks

As we discuss shortly, all this communication within the home takes place over some sort of network, either wired or wireless. Most home automation today is done wirelessly, using Wi-Fi, Bluetooth, or similar proprietary networking technologies.

Smarter Living with Smart Furniture

Let's start our examination of individual items in the smart home with something that we all have in our homes—furniture. Now, it's hard to think of your couch or coffee table as being exceptionally smart, but where there's a will, there's a way.

South Korean furniture maker Hyundai Livart has partnered with that country's largest mobile carrier, SK Telecom, to combine smart home technology with traditional household furniture. Each item in the smart furniture line has its own built-in touchscreen that replicates the owner's smartphone display via mirroring technology. For example, the company offers a dresser with touchscreen displays in the cabinet doors, as well as mirrors with touchscreens built in (shown in Figure 5.1).

Figure 5.1   *The touchscreen display embedded into a smart mirror from Hyundai Livart/SKT.*

People can use these smart furniture items to surf the Internet, listen to streaming radio, display news and weather, search for recipes, and the like. Since we're talking smartphone mirroring, the furniture can also be used to make and receive phone calls. Future functionality will enable users to control room temperature, open and lock doors throughout the house, and monitor visitors via a front-door camera.

Or how about the ReST bed, from Responsive Surface Technology? This smart bed incorporates special fabric and sensors to track your sleep habits, and then uses that data to keep you comfortable when you toss and turn all night. The ReST bed features 18 air sensors that inflate and deflate according to your preferences and your nocturnal activity.

Then there's Luna, a smart mattress cover you can add to any bed. It's packed with sensors to monitor your sleep patterns, and then communicates the data it collects to a variety of other smart devices. For example, Luna can sense when you're about to wake up in the morning, and then instruct your smart coffee maker to start making coffee, tell your smart thermostat to crank up the heat, and even have your smart lighting system light up to brighten your morning. (Conversely, it can tell that same lighting system to turn off any still-lit light when you fall asleep—as well as make sure your smart locks are all engaged.) Luna can also analyze your sleep cycle, heart rate, and breathing rate to adjust its temperature in real time to improve your sleep experience.

Smarter Environment with Smart Lighting

Here's some pretty popular smart home technology that's readily available today. Smart lighting systems enable you to control the lights inside and outside your house. Lights can be controlled on a preprogrammed time cycle or configured to automatically shut off when a room is unoccupied—but turn back on when someone walks into the room.

These systems typically require the smart bulbs be plugged into some sort of physical bridge or gateway device. You then connect to the gateway via the system remote control or a smartphone app and send the necessary commands to all connected bulbs. (Figure 5.2 shows how it works with the Connected by TCP system.) For that matter, you can use the system to control the lighting in your house while you're away, over the Internet.

Even neater, smart LED lights can be configured to output a specific brightness or color. This lets you provide different levels and types of lighting for different tasks.

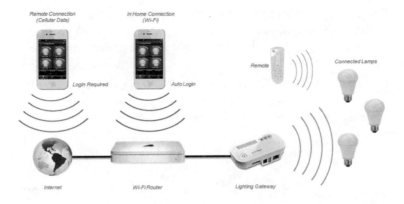

Figure 5.2 *The Connected by TCP smart lighting system.*

 Note

One of the advantages of LED lighting is that the individual LEDs in a bulb can be adjusted to different brightness levels and different colors. You can't do that with compact fluorescent lamp (CFL) or traditional incandescent bulbs.

There are a number of companies offering smart lights and smart lighting control systems. These include:

- **Belkin**—Its WeMo Smart LED Bulbs can be controlled by the company's WeMo Link system. Insert a WeMo bulb into your lamp and then control it (over Wi-Fi or 3G/4G) with the WeMo Light Switch, WeMo Motion (motion detector), or WeMo App. The WeMo App lets you register each light bulb, rename it, and customize rules for its operation.

- **FX Luminaire**—Specializes in landscape and architectural lighting. The company offers a variety of LED lights that can be controlled via traditional programming or smartphone app.

- **Lutron**—Has long been a player in custom-installed, whole-house automated lighting systems. The company offers a variety of home automation solutions that enable you to control your home's lighting, window coverings, and heating/cooling systems—all wirelessly, using Lutron's controllers or smartphone apps.

- **Philips**—Offers Hue Connected Bulbs that can be controlled via the company's smartphone app, shown in Figure 5.3. You can program individual LED bulbs to turn on or off, brighten or dim, or change color.

- **Connected by TCP**—Offers smart LED bulbs of various types and sizes that, when connected to the company's lighting gateway, can be controlled by the Connected remote control or their smartphone app.

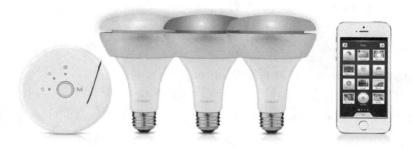

Figure 5.3   *Philips' Hue Connected Bulbs, bridge controller, and smartphone app.*

These smart lights and lighting systems are not cheap. For example, Philips offers a starter pack for its Hue Connected system with three bulbs and a bridge controller that sells for around $200. TCP offers a similar Connected starter pack with two bulbs and a gateway for $80. Individual smart LED bulbs sell for $25 or more apiece.

Smarter Views with Smart Windows

Lutron, one of the big players in smart lighting, is also into the smart windows market. Actually, it's not the windows that are smart, but rather the window coverings, in the form of automatically controlled blinds and curtains.

Motorized Window Coverings

With smart windows, your window coverings are mechanized, meaning they're connected to motors that can open and close them. You then control the motors via a timer or other programmed controller, the system's own remote control unit, or a smartphone or computer app. This way you can have your curtains open and close just like you're home, even if you're not. Plus, closing or opening all your drapes and blinds for a room or your whole house is as easy as pressing a button on your phone.

In addition to Lutron, motorized window treatments are offered by a variety of companies, including Bali, HunterDouglas, Serena, Somfy, and other companies. (Figure 5.4 shows a typical motorized blinds system, from HunterDouglas.)

Figure 5.4 *A motorized blinds system, complete with wall control and remote control, from HunterDouglas.*

Smart Glass

In the future, you might not need curtains and blinds, because the window glass itself will become smarter. We're talking smart glass that changes from clear to tinted on command and selectively blocks light, glare, and even heat from the outside. It works by adding a reconfigurable layer within or on the outside of the glass. Some companies use a ceramic coating that darkens when low-voltage electricity is applied; other companies are working with tin oxide nanocrystals.

Smart glass technology is probably a decade away from the consumer market and is likely to cost twice as much as traditional windows. But given that you won't need to buy curtains—and the potential energy savings—it may represent a net savings.

Note

Smart glass is also called *dynamic* or *switchable glass*. Companies working on the concept include SageGlass and View, the latter of which is backed by Corning, the big glass manufacturer.

Smarter Heating and Cooling with Smart Thermostats

Just about every home today has some sort of thermostat, which serves as the controller for your furnace and air conditioner. The smarter the controller, the smarter your heating and cooling.

Nest Learning Thermostat

One of today's more publicized smart devices, Nest's Learning Thermostat not only learns how your family wants your home heated or cooled and adjusts itself accordingly, but reports back to Nest about these habits. The company then furnishes this data to your local utility company—which we expand upon in a moment.

As you can see in Figure 5.5, the Nest thermostat is a nicely designed little controller, much more attractive than your typical block-like programmable thermostat. It's a sleek metallic dial with round liquid crystal display (LCD) that looks more like an Apple product than the industrial equipment you find at your local Home Depot. The display goes orange when the heat is on and displays a cool blue when you're using the A/C. Operation is simplicity itself; turn the dial to the right to turn up the heat or to the left to cool things down.

What's neat about the Nest thermostat is that it learns from your interaction. After a bit of time, it knows that you turn up the heat when you wake up on a winter morning, or turn down the air when it gets warm on summer afternoons. All it takes is a week or so, and it knows all your habits—and adjusts itself accordingly. It even knows when you leave the house, so it can put itself into energy-saving Away mode.

Speaking of being away, you can easily monitor and control the Nest thermostat from wherever you are, via the Nest Mobile smartphone app, shown in Figure 5.6. Want a hot house cool when you arrive home from work? Just make it so via the Nest Mobile app.

Figure 5.5 *The Nest Learning Thermostat.*

Figure 5.6 *Controlling your home's heating and cooling via the Nest Mobile app.*

The Nest connects via Wi-Fi to your home network, and then to your furnace and air conditioner. It also uses Wi-Fi to connect to your home computer for various reports and diagnostics, and to the Nest mother ship to transmit all your usage data.

As to those reports and diagnostics, you can use your computer or tablet to review your energy usage via Nest. You can examine your energy history and see how much money Nest saved you in the previous days and weeks.

Of course, the Nest costs more than a typical programmable thermostat—$249 versus $50 or so for traditional thermostats. Nest says you'll save more than the difference because of more efficient energy usage, and that's likely. Here's a smart appliance that has real discernable consumer benefits.

 Note

Nest has proven so popular that the parent company, Nest Labs, was recently acquired by Google, for a cool $3.2 billion. Wowzers.

Other Smart Thermostats

Nest isn't the only touchscreen learning thermostat, of course. Honeywell, for example, offers its own Wi-Fi Smart Thermostat. As you can see in Figure 5.7, this unit looks a bit more conventional than the Nest dial but offers much of the same automation functionality. It also displays your local weather conditions on the large LCD screen.

Figure 5.7   *Honeywell's Wi-Fi Smart Thermostat.*

Honeywell also offers a Nest-like Wi-Fi thermostat dubbed the Lyric Thermostat. As you can see in Figure 5.8, it's a circular unit, like the Nest, and offers app-based operation. It also sends messages to its smartphone app to alert you to upcoming extreme temperatures or the need to change filters.

Figure 5.8 *Honeywell's Lyric Thermostat and smartphone app.*

Also playing in this space is Ecobee, with its Ecobee3 Smart Thermostat, shown in Figure 5.9. This unit looks and works a lot like the Nest model; it's Wi-Fi–enabled and can be controlled via smartphone app. What's unique about the Ecobee is that it uses motion and proximity sensors to know when you're home or away. It costs about the same as the Nest.

Figure 5.9 *The Ecobee3 Smart Thermostat and room sensor.*

Note

In the future, you'll find more smart thermostats that let you automatically control your home's temperature and humidity, wirelessly and remotely. More advanced systems will incorporate control of your smart window coverings or smart windows, even to the point of automatically opening and closing selected windows to cool your home on moderate days.

Using Nest with Other Smart Devices

Nest's wireless connectivity enables it to share data and interact with other smart devices in your home. For example, Whirlpool offers a washer/dryer combo that connects to the Nest thermostat (via Wi-Fi, of course) to enable more efficient operation. In this instance, the Nest thermostat tells Whirlpool's Smart Front Load Washer and Dryer when you're away. The washer and dryer, shown in Figure 5.10, then adapt their operation to use longer (that is, cooler) dryer cycles to save energy, and to keep your clothes tumbling so that they stay fresh and unwrinkled until you get home. The Whirlpool equipment can also share information from the Nest about energy rush hours and auto-delay laundry cycles accordingly.

Figure 5.10 *Whirlpool's Smart Front Load Washer and Dryer connect to and work with the Nest thermostat.*

Nest's interactivity goes beyond the laundry room. At present, Nest works with the following smart devices:

- **Chamberlain**—Connect your Nest thermostat to your Chamberlain garage door opener. Close your garage door when you go out and the garage door opener tells the Nest thermostat so it can activate Away mode. Same thing in reverse when you get home, too.

- **Jawbone**—Use your Nest thermostat with Jawbone's UP24 wearable device, shown in Figure 5.11, and Nest will know when you're awake or asleep and adjust the temperature accordingly.

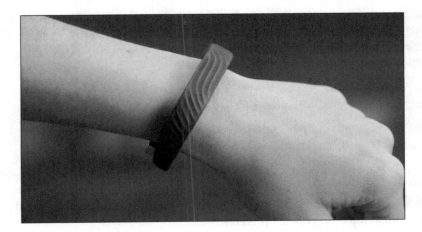

Figure 5.11 *Jawbone's UP24 activity tracking band tells Nest when you're awake or asleep.*

- **LIFX**—When the Nest thermostat switches into Away mode, LIFX smart lighting automatically turns on or off lights throughout your house to make it look as if you're at home.

- **Logitech**—Add Nest thermostat controls to the Activity screen of the Logitech Harmony remote, so that when you start an activity (such as watching a movie or listening to music), you adjust the room temperature as well as turning on the proper equipment.

- **Mercedes-Benz**—The ultimate smart connection. Your new Mercedes can alert your Nest thermostat when you're close to home so it can turn up (or down) the temperature accordingly.

And there's more to come.

Data Collection and Control Issues

Now a word about all that data that the Nest thermostat collects—basically, how much energy you're using and when. Nest deals with most major electricity companies to provide this data from its users (and you agree to this when you blindly check Okay to the thermostat's terms of use).

Nest says that it doesn't share your actual data with your electric company; it only reports data in aggregate. That may be small consolation, however, when you discover the other part of the deal.

You see, Nest's deals with these utility companies give them the option of remotely turning up your A/C on a hot day or when the power load is sufficiently heavy. (Actually, it's Nest that turns up your thermostat, based on requests from your power company.) This enables the power company to level out the load on its grid, which saves it money in the long haul.

So you cede control over your thermostat, based on the needs of your power company. And guess who profits from that? It's not you, bub.

The power companies that Nest partners with pay Nest somewhere between $20 to $50 per thermostat per year. You don't see a penny of that.

You also don't see a penny of the savings that your utility company realizes from this whole operation. Nest, however, gets to split the savings with your power company. For every dollar the power company saves based on the thermostat you purchased, Nest takes home fifty cents.

This ends up being quite lucrative for Nest. The company makes money when you initially pony up two and a half bills for your thermostat, and then every month afterwards from the power companies. According to *Forbes* magazine, which first reported this, Nest's revenue from utility companies outweighs revenue from the sales of those thermostats.

So as much as you might like how the Nest thermostat looks and how it works, Nest isn't doing all this for the ultimate benefit of mankind. That cool-looking thermostat glowing blue on your wall is just another way for a big company to make big money at your expense. Yeah, you might save a little on energy costs, but Nest is making more by supplying your data—and your tacit control—to the big energy companies. Consider that before you buy.

Smarter Protection with Smart Security Systems

Many homeowners already have a bit of home automation installed, in the form of a home security system. Today's home security systems incorporate a variety of motion sensors, proximity sensors, and door/window sensors, all connected to a main control unit. When the system is switched on and one of the sensors is breached, a signal is sent to the control unit. This may sound an alarm, turn on some lights, or send another signal to the system's monitoring company. The company will then place a 911 call to local police and then things really start jumping.

Smarter Security Systems

More intelligent home security systems enable you to activate the system on a room-by-room or zone-by-zone basis. Some systems also integrate with smoke alarms and carbon monoxide sensors, to alert the monitoring company when there's a fire or deadly gas in the house.

Things get more interesting when you connect your home security system with other home automation systems. For example, pressing a single button might lock all outside doors, arm the alarm system, close the motorized curtains, turn on selective lighting, and turn down the furnace.

Smart Locks

Of course, for a smart security system to be able to lock your doors, you need some sort of smart lock. These exist. The smart lock (sometimes called a *connected lock*) installs in place of your existing door lock and connects to your home network via Bluetooth or Wi-Fi. You can then operate the lock with a smartphone app, which means you no longer need your keys to open the front door. Smartphone operation also means you can lock or unlock your doors while you're away from home; future smart locks will also be connected to your whole-house smart security system. Smart locks are currently available from August, Goji (shown in Figure 5.12), Kwikset, Lockitron, and Schlage.

Smart Security Cameras

In addition, expect most future smart security systems to include a bevy of cameras mounted throughout your house and around your property. You can view the cameras from your smartphone or notebook personal computer (PC), wherever you may happen to be, over the Internet.

One fun little camera you can install today is the SkyBell, shown in Figure 5.13. The SkyBell is billed as a smart video doorbell, as it includes a built-in video camera. The doorbell interfaces via Wi-Fi to the accompanying smartphone app, so you can see who's at the door on your phone—even if you're away on vacation. Similar smart doorbells are available from Chui, i-Bell, and Ring.

Figure 5.12   *Protect your home with the Goji Smart Lock.*

Figure 5.13   *View who's ringing your bell with the SkyBell smart doorbell.*

Smarter Sensing with Smart Monitors

Home security systems help protect you from unwanted visitors. But there are more potential dangers lurking in your home. What if a water line breaks? What if there's carbon monoxide gas in the house? What if there's a fire?

To protect against all these dangers—and more—a variety of smart monitors are being introduced to the market. These devices monitor for specific dangers, and alert you, typically via a smartphone app, when something bad is detected.

Envision this scenario. You're away from home, on vacation or a business trip, and your smart monitors have all been activated. The moisture sensor in your basement detects water on the basement floor, while the sensor attached to your water line detects a higher-than-normal amount of water flow. The information from both sensors is sent to a central monitoring device, which puts two and two together to guess that you have a water leak. The main device notifies you via the convenient smartphone app that you have an issue, and also notifies your next-door neighbor or other primary contact, as well as your designated plumbing service. The neighbor and plumber are sent entry codes to let themselves into your house, via your smart door lock. You're notified, also via the smartphone app, when each of these people enter your house and again when they leave. The plumber does his thing, the smart devices reset, and the bill for his services arrives electronically in your inbox.

Smart Smoke Detectors

Other smart monitors operate more directly when activated. Just as today's dumb smoke alarms beep when smoke is detected, tomorrow's smart alarms will still beep but also alert you via smartphone app, as well as dial your local fire department with the alert.

This type of smart smoke detector exists today, in the form of the Nest Protect, shown in Figure 5.14. The Nest Protect is a combination smoke and carbon monoxide detector. Like the Nest Learning Thermostat, it's a smart monitor that connects to the accompanying smartphone app via Wi-Fi. At the first indication of any problem, you get a friendly heads-up on your smartphone. If things get worse, the Nest Protect flashes red, sounds an alarm, and tells your household what to do, using recorded words instead of the normal beeps. So you might hear the alarm say "There's smoke in the living room," if that's what the problem is.

Nest Protect also works in conjunction with the Nest thermostat. If Nest Protect detects high carbon monoxide levels, it notifies the Nest thermostat, which then turns off your gas furnace. Pretty nifty.

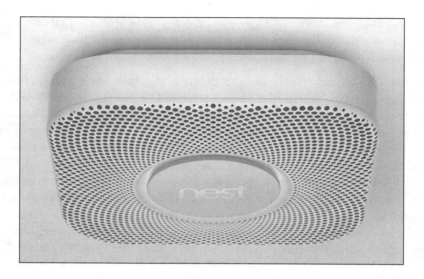

Figure 5.14   *The Nest Protect smart smoke and carbon monoxide detector.*

 Note

Normal smoke detectors can be had for $10 to $20 or so. The Nest Protect costs more than that, obviously. At $99 each, the Nest Protect seems mighty expensive, especially if you install several throughout your home. But that's the way it is with smart technology today—you pay more to be on the cutting edge.

Smart Air Quality Monitors

Then there's Birdi, a smart air quality monitor that acts much like the proverbial canary in the coalmine. In addition to monitoring carbon monoxide levels, it measures temperature, humidity, dust, soot, pollen, particulate levels, and more. It can tell you if your allergies are likely to act up or if someone is smoking in another room. It also acts as a smoke detector, so there's that. Whatever the issue is, you get alerted via the accompanying smartphone app, shown in Figure 5.15, in plain-English messages. Birdi will even signal you if there's some sort of natural disaster in the making, such as a tornado or earthquake.

Even better, Birdi alerts your neighbors if something's wrong at your house. (And vice versa, if they have Birdi units installed.) If there's any sort of emergency, fire or gas or whatever, Birdi will call your neighbors and provide a recorded notification. It's kind of a one-stop-shop smart monitor, priced at $119.

Figure 5.15 *Monitoring air quality in your home with Birdi's smartphone app.*

Smarter Information with Amazon Echo

There's a new smart device on the market that doesn't easily fall into any of our previous categories. The device comes from Amazon, the online retailer, and it's called Echo. It's kind of a smart brain for your home.

You know how iPhone's Siri works. Well, Amazon's Echo works kind of the same way. It's an always-on device that you communicate with via spoken language. Wake it up by saying "Alexa" (that's the device's artificial personality), and then ask it something. You can use Echo to find the latest news and weather reports, find information on the web, and (of course) purchase items from Amazon. Ask Echo, "How many teaspoons in a tablespoon?," "Will it rain tomorrow?," or "How old is George Clooney?" It'll tell you the answers.

Now, Echo itself doesn't have all the answers built in; it's not in and of itself an intelligent device. Instead, Echo connects to the cloud (via your home Wi-Fi network) to find information there. You can also use Echo to set timers and alarms, create shopping and to-do lists, and such.

The Echo unit has a built-in speaker system, so you can use it to play music from Amazon Music, iHeartRadio, and TuneIn Radio. Echo is also Bluetooth-enabled, so you can stream Spotify, Pandora, and iTunes music from your smartphone or tablet. And, as you can see in Figure 5.16, it's small enough to sit on a kitchen counter.

Figure 5.16   *Get smart answers from the Amazon Echo.*

I'm not sure we really need another freestanding device that essentially duplicates the functionality built into today's smartphones and tablets. That said, the Echo is an interesting first step towards the fifth level of smart home technology we discussed earlier in this chapter. It sells for $199.

Reimagining the Smart Network

To operate all the smart devices in your smart home, and for them to communicate with one another (and with you), all those devices need to be connected to some sort of network. Today, that's likely to be your home's wireless Wi-Fi network, which then enables you to interface with them via your smartphone, tablet, or PC.

Not everyone thinks that Wi-Fi is the best way to connect your smart devices. Some devices connect via Ethernet instead, which is a bit unwieldy unless you absolutely, positively need that hard-wired connection. Bluetooth is used sporadically—typically to connect to other devices that then connect to your Wi-Fi network.

In the home automation market, three companies or consortiums have introduced their own wireless networking technologies that have meet with widespread approval among both professional installers and consumers. These technologies are INSTEON, Z-Wave, and ZigBee.

All three of these technologies utilize wireless mesh networks. As you learned in Chapter 2, "Smart Technology: How the Internet of Things Works," mesh networks work by connecting one device to another device and then to another. This creates a "mesh" of connections, unlike the hub-and-spoke model used by Wi-Fi and other traditional networks. Each device in the mesh network acts as a repeater, receiving and sending every message to all other connected devices. The more devices added to the network, the stronger it becomes.

 Note

> The mesh approach is thought by some to be superior for use by the multiple simple devices found in smart homes.

It goes without saying that a device that works on a Z-Wave network won't be able to connect to another device that works with ZigBee. Unless, that is, each device connects to a master hub that incorporates both Z-Wave and ZigBee technologies. As you've seen, several companies offer just such dual- or tri-protocol hubs; most also offer Ethernet or Wi-Fi connectivity so the whole shebang can connect to the router on your home network.

Let's take a quick look at these three competing smart networking technologies.

INSTEON

INSTEON is a leading developer of networking technology for the connected home. The company's devices connect via a combination of wireless radio frequency (RF) and wired power line technologies. This two-layer network enables control signals to jump from one layer to another if problems are encountered, thus enhancing both speed and reliability.

As noted, INSTEON's RF networking operates in a mesh configuration. Because of the way mesh networks work, INSTEON systems have no need for a central controller. Setting up an INSTEON system is as simple as plugging each device into a special power connector. This automatically connects the new devices to all existing devices in your home.

In addition to supplying the underlying technology, INSTEON sells a variety of different smart devices that connect to their mesh networks. These include light switches and dimmers, smart light bulbs, motion sensors, thermostats, Wi-Fi cameras, and sprinkler system controls. These devices can be controlled from INSTEON's own remote controls, or from the INSTEON iOS and Android apps.

Note that unlike Z-Wave and ZigBee, INSTEON has not been made available to or been widely adopted by other home automation companies. It's more of a closed system.

Z-Wave

Z-Wave, on the other hand, is a wireless technology used by many home automation suppliers. The company claims that more than 20 million Z-Wave-compatible devices have been sold to date, which makes it one of the, if not *the*, most popular technologies for connecting together smart devices in the home. The technology is popular both with home DIYers and professional installers.

Z-Wave networks operate in the 900MHz band, so there's no interference with 2.4GHz Wi-Fi networks. It's a low-energy technology, so Z-Wave devices can operate on battery power alone (although many devices plug into the wall for power). A single Z-Wave network can handle up to 232 devices, with an average maximum range between devices of 300 feet or so.

ZigBee

ZigBee works similarly to Z-Wave and is supported by a similar number of home automation companies. ZigBee networks operate in the 915MHz band (in the U.S., anyway; other bands are utilized in other countries), so there's no competition with either Z-Wave or Wi-Fi networks. The mesh configuration enables a 30 to 60 foot connection range between devices.

The ZigBee Alliance claims that there are more than 1,000 ZigBee-compatible products on the market, from 400 different manufacturers. It's particularly popular among industrial suppliers; for example, ZigBee technology is incorporated into tens of millions of smart gas and electricity meters worldwide. ZigBee is also big with DIYers, who use the technology for all manner of DIY projects.

Controlling the Smart Home

Individual smart devices can be extremely useful in and of themselves, but to create a truly smart home, you have to tie all these devices together. To that end, several manufacturers have smart hubs and controllers to operate the various smart devices that they offer. These are *smart home systems*, and they're all relatively proprietary—that is, you can't operate one manufacturer's devices with another company's smart hub. Pick a system and then stay with it.

There are a number of these smart systems out there. Let's take a quick look at the most popular.

Control4

Control4 is a manufacturer of home automation equipment for the installer market. Its product line revolves around a Linux-based hub dubbed the Home

Controller. This device controls all the connected devices in the home—both the company's own products and devices from other companies.

The Control4 concept focuses on controlling entire rooms rather than the equipment within a room. A "room" can contain devices physically located in another room, such as smart window coverings, thermostats, audio/video systems, and the like. Select a room and then select a scene for that room. For example, you might select Living Room and then Watch DVD. At this point, the Control4 system turns on the TV in that room, switches it to the DVD input, turns on the audio/video receiver, switches it to the appropriate input, dims the lights, closes the drapes, and presses Play on the DVD player. When you're done watching, press a single Room Off button and everything switches off and returns to its previous state.

Control4 supports a number of smart devices from other manufacturers and also offers its own line of devices. These include touchscreen controllers, wall keypads, multi-zone audio amplifiers, light switches and dimmers, smart thermostats, and the like. To communicate with connected devices, the Control4 system uses either TCP/IP over the Internet or ZigBee's wireless mesh networking technology.

Crestron

Crestron is a long-standing leader in home automation systems. Its products are widely used by home theater and whole-home systems integrators; we're talking high-end stuff here.

Crestron offers a variety of smart home devices, to control TV and music playback, lighting, window treatments, temperature, door locks, and other home security. With everything connected to Crestron's main hub, you can press one button on a remote or smartphone app to enable a specific "scene."

For example, enable the Wake scene in the morning and your bedroom lights will slowly ramp up to the desired level, the lights in your closet and bathroom turn on, the TV built into the bathroom mirror turns on and changes to your favorite news channel, your favorite music plays softly in the background, and the heated bathroom floor warms up. Enable the Watch scene in your home theater room and the shades are lowered, the lights dim, and the movie starts playing with the volume set to the perfect level. It's a one-button world.

Like Control4, Crestron definitely plays at the high end of the smart home market. Its products are designed for professional installers doing whole-house systems; it's not stuff you can pick up cheap at your local Home Depot. Still, if you want a system that just works, no fuss and muss, Crestron is the way to go.

HomeSeer

HomeSeer offers a full line of home automation controllers, starting at just $199. HomeSeer's HS3 home automation software is embedded in all the company's controllers, and works (via plugins) with a large number of home automation products and technologies, including INSTEON, X-10, and Z-Wave.

You use HomeSeer controllers to control smart devices from other companies. With the appropriate HomeSeer controller(s), remote control devices (such as the Control Pad Tabletop Touchscreen, shown in Figure 5.17), and smartphone apps, you can control your home's temperature, lighting, audio/video playback, door locks, garage doors, webcams, and security system.

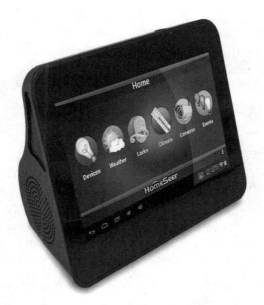

Figure 5.17 *Control devices connected to a HomeSeer controller with the HomeSeer Control Pad Tabletop Touchscreen.*

Iris

Iris is a line of home automation devices sold by Lowes, the big home improvement retailer. As such, it's designed with the average consumer in mind—in terms of both ease-of-install and price.

Lowes offers Iris control hubs, smart wall plugs, contact sensors, motion sensors, thermostats, surveillance cameras, smoke detectors, water leak detectors, door locks, and more. Just connect a lamp or other electrical device to the smart wall plug, and it can be controlled by the Iris system.

You can control Iris via the obligatory smartphone app or by voice commands when you're in the house. If an alarm is triggered, you can configure Iris to send you email, text, or voice messages.

Iris components are controlled by a central hub, which connects to your home Wi-Fi network. The hub can also receive commands from Z-Wave devices and contains a built-in speaker that sounds alarms and announcements.

The unusual thing about Iris is that, to take full advantage of its functionality, you need to subscribe to the Iris monitoring service. This runs $9.99/month and essentially serves as the gateway for all communication between your Iris devices and yourself.

mControl

Embedded Automations has developed its own digital home software control platform, dubbed mControl. This platform connects to devices using INSTEON, X10, and Z-Wave technologies and is used to control connected lighting systems, thermostats, home security systems, audio/video systems, and more. The company doesn't sell its own devices, only the controller software which can then control devices from other companies.

Quirky

Quirky is a startup focusing on smart inventions, and they've partnered with behemoth General Electric (GE) to release a line of smart home devices. Each device is Wi-Fi–enabled and can be controlled via iOS or Android app. Its more interesting devices include the Spotter UNIQ, which combines light, sound, motion, and temperature sensing in a single device; the Overflow water leak sensor; the Egg Minder, which keeps tracks of egg freshness in your fridge; the Ascend garage door opener controller; the Tapt smart wall switch; and Norm, a kind of smart thermostat.

SmartThings

SmartThings is an interesting company—so interesting it was recently acquired by Samsung, which has big designs on the IoT market. SmartThings plays in the consumer end of things, which means affordable and easy-to-install kits and individual devices.

Everything in the SmartThings system is controlled by the $99 SmartThings Hub, which you can control from the SmartThings app, shown in Figure 5.18. The company offers a variety of different sensors and devices, including temperature and humidity sensors, motion sensors, moisture sensors, and smart power outlets.

Figure 5.18   *Controlling your smart home with the SmartThings smartphone app.*

The SmartThings Hub is compatible with both ZigBee and Z-Wave wireless technologies. It doesn't use Wi-Fi, although you can connect it to your home router via Ethernet. (And you need to if you want to enable Wi-Fi smartphone control.) Because of the ZigBee/Z-Wave connectivity, you can use the SmartThings Hub with devices from many different third-party suppliers.

Vera

Vera also plays in the consumer end of the home automation market. Everything runs through one of several different smart controllers, which interface to other devices via Z-Wave wireless and to your home network via Ethernet or Wi-Fi. As such, Vera can control other Z-Wave-compatible devices, as well as the company's own video cameras, dimmable lamp modules, motion sensors, and the like.

Vivint

Vivint is another higher-end supplier of home automation systems. The company offers home security, heating/cooling, and control solutions. It's the type of system typically offered by professional installers.

WeMo

Back into the consumer end of things, we have Belkin's WeMo home automation products. WeMo is one of the more popular systems among DIYers; the products work together well, are easy enough to install, and easy enough to operate.

WeMo devices work via Wi-Fi, which is convenient for most home users. They also work with IFTTT (if this, then that), a programming protocol which uses conditional statements to initiate specific operations.

What's especially great about WeMo is the variety of smart devices offered. We're talking smart LED bulbs, light switches, power switches, and webcams. Even better, Belkin is partnering with other companies to include WeMo technology in coffee makers (by Mr. Coffee), slow cookers (by Crock-Pot), and humidifiers (by Holmes). For example, the WeMo Crock-Pot, shown in Figure 5.19, enables you to turn it on remotely via the WeMo smartphone app so you can start cooking while you're still at work.

Figure 5.19 *Operate this Crock-Pot slow cooker via the WeMo app.*

Wink

Wink is an offshoot of Quirky, with controllers for Quirky devices and devices from other companies. There's the Wink Hub, Wink Relay touchscreen controller (shown in Figure 5.20), and Wink smartphone app. You can use either controller to operate devices connected to the Wink Hub.

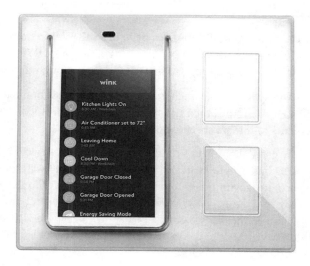

Figure 5.20   *Use the Wink Relay touchscreen controller to operate smart devices from a number of different companies.*

Devices can connect to the Wink Hub using Bluetooth, ZigBee, or Z-Wave wireless technologies. The Hub also works with proprietary technologies used by Kidde and Lutron. This means you can use the Wink Hub to control smart devices from a number of third-party companies, which is a good thing.

X10

When it comes to home automation systems, X10 is the granddaddy of them all. X10 has been around since the mid-1970s and long has been a favorite among DIY consumers.

The X10 system operates over your home's existing power lines. Just connect an electrical device to an X10 Appliance Module, which then plugs into a nearby wall outlet. With a similar X10 Transceiver Module plugged into another outlet, use the X10 Controller to wirelessly operate any plugged-in device.

Admittedly, X10 is kind of old tech in our new tech world. That said, it works, and that's saying something. It's a low-cost way to dip your toes into the world of home automation.

 Note

> There are also open source home automation/IoT controllers, popular with the DIY crowd. Because the underlying code is openly available to anyone, these controllers are easily hacked or modified, which makes them ideal for special projects and customization. One such open-source home automation controller is the Thing System from Alasdair Allan and Marshall T. Rose; learn more at www.thethingsystem.com.

Where Can You Find Smart Home Devices?

We've talked about a lot of smart home technologies and devices in this chapter. If you're interested in any of these products or companies, here's where you can find more information:

- Amazon (Echo), www.amazon.com/echo
- August Smart Lock, www.august.com
- Bali, www.baliblinds.com
- Belkin, www.belkin.com
- WeMo, www.wemothat.com
- Birdi, www.getbirdi.com
- Chui, www.getchui.com
- Control4, www.control4.com
- Crestron, www.crestron.com
- Ecobee, www.ecobee.com
- Embedded Automation (mHome and mControl), www.embedded-automation.com
- FX Luminaire, www.fxl.com
- Goji, www.gojiaccess.com
- HomeSeer, www.homeseer.com
- Honeywell, yourhome.honeywell.com
- HunterDouglas, www.hunterdouglas.com
- i-Bell, www.i-bell.co.uk
- INSTEON, www.insteon.com

- iRobot (Roomba), www.irobot.com
- Kwikset, www.kwikset.com
- Lockitron, www.lockitron.com
- Lowes (Iris), www.lowes.com/iris
- Luna, www.lunasleep.com
- Lutron, www.lutron.com
- Nest, www.nest.com
- Philips (Hue lighting), www2.meethue.com
- Quirky, www.quirky.com
- Responsive Surface Technology (ReST), www.restperformance.com
- Ring, www.ring.com
- SageGlass, www.sageglass.com
- Schlage, www.schlage.com
- Serena, www.serenashades.com
- SK Telecom (smart furniture), www.sktelecom.com
- SkyBell, www.skybell.com
- SmartThings, www.smartthings.com
- Somfy, www.somfy.com
- TCP (Connected lighting), go.tcpi.com/GetConnected
- Vera, www.getvera.com
- View, www.viewglass.com
- Vivint, www.vivint.com
- WeMo, www.wemothat.com
- Wink, www.wink.com
- X10, www.x10.com
- ZigBee, www.zigbee.org
- Z-Wave, www.z-wave.com

SMART HOMES AND YOU

Whew. All this talk about smart homes and smart devices can be exhausting. There are a ton of different devices to choose from, and more on the way, all using different controllers and networking technologies. How do you decide what products—if any—to use in your home?

If you want to test the smart home waters, investing in some smart lighting is probably the easiest (and lowest-cost) way to do so. Buy a smart lighting kit from Philips or Belkin and play around with that awhile. If you like what you get, then buy some more smart bulbs and keep going from there.

At the next level, consider buying a smart thermostat. While there are several worthy products available in this category, you can't go wrong with the Nest Learning Thermostat. Nest is a proven product at this point, and the company has connections with several other companies that enable their smart devices to connect to the Nest thermostat. It's a little pricey, but it works very well and you can get a taste of what smart home technology can do. (And if you like the way the thermostat works, you can always supplement it with the Nest Protect smart smoke detector.)

Beyond that, you can go whole hog and start adding other smart devices from other companies. We're talking smart power switches, smart sensors, and the like. At this point, you probably want to make a decision about which controllers and networking protocols you want to support. Belkin's WeMo system is nice in that it uses common Wi-Fi connections. Otherwise, look for a controller or system that is compatible with both Z-Wave and ZigBee products, so you have more choices available to you.

Once you dip your toes into the waters, you'll start to see the potential offered by smart home technology. What we have today is the proverbial tip of the proverbial iceberg; keep your eye on ongoing developments to see what's coming down the highway in the future.

6

Smart Clothing: Wearable Tech

We already have wearable fitness tech that helps you monitor your exercise and workouts. Apple and other companies are working on smartwatches to supplement or replace much of the functionality of smartphones. What smart clothing is next?

A world of wearable wonders awaits at the conjunction of the high tech and clothing industries. You will soon be wearing the bastard children of this unlikely alliance— for better or worse.

Wearable Technology Today—and Tomorrow

Wearable tech describes a wide range of devices embedded or integrated into all manner of clothing and accessories. It might seem strange to be talking about smart shirts and smart socks, but when just about everything, no matter how small, can have an IP address and Bluetooth transmitter, the sky's the limit.

What might be surprising is just how much of this new high-tech clothing is available today. While wearable tech might not be mainstream enough to occupy space in the men's clothing department at Target, it's inching closer. (In fact, it's a big enough business that Amazon has its own Wearable Technology store online.)

The promise of wearable technology is to bring you the advanced functionality found in a typical smartphone or computer app, but incorporated into the things you wear every day. We're talking the ability to collect, analyze, and display useful and important data without having to pull out your smartphone or notebook PC. It's data you use, at a glance, keeping your hands free.

Technology-enabled clothing and accessories are made possible because of ongoing advancements in miniaturization. Chips are smaller, sensors are smaller, transmitters are smaller, even displays are smaller than ever before, which means it's relatively easy to integrate the necessary technology into the form factor of a watch, wristband, or t-shirt. It also helps that technology costs keep decreasing, so that these nifty gadgets continue moving down into the near-affordable range. A smart t-shirt is still going to be more expensive than a plain white tee, but not that much more expensive than the fancy training shirts you find in your favorite sporting goods stores.

Now, few of these wearable devices do more than what you can accomplish with your handy dandy iPhone, but that's not the point. The point is to provide this sort of targeted usability in a smaller form factor, so that you don't always have to be whipping out the ever larger smartphone. It's application-specific stuff built into clothing or accessories you can wear every day.

Interestingly, the companies that make these wearable devices not only have to deal with the technology, but also with the fashion. It's not good enough to produce a functional smartwatch or fitness band; the thing has to look as good as it works. Clothing isn't all about functionality, after all. It's also about fashion and comfort. That's a challenge to tech companies such as Apple and Microsoft, but one their designers are no doubt aware of.

Certainly, there is sufficient incentive for these companies (and lots of new start-ups) to get it right. Experts predict wearable technology will become a $70 billion business over the next decade, so there's lots of money to go around. The biggest segment of this business will center around health-related devices, such as fitness trackers, medical devices, and the like. But smartwatches also represent a big opportunity, as do completely new types of devices, such as Google Glass and other heads-up displays.

Watching the Smartwatches

When most people today think about wearable tech, they first think of so-called smartwatches. That's not because there are a lot of them being sold (there aren't—at least, not yet), but rather because this category gets written about a lot in the media. Smartwatches are sexy.

It's obvious that the smartwatch, like all wearable tech, is an emerging product. Only 2 million smartwatches were sold in 2013, although Business Insider forecasts that by 2018 more than 90 million of the things will be sold each year. If that prediction comes true, and figuring an average selling price of around $200 (which may be low), that translates into an $18 billion market, which ain't small potatoes. (And explains why Apple wants a piece of the pie.)

What exactly is a smartwatch? It all depends—a smartwatch can be different things to different people. Some smartwatches are nothing more than watches with built-in digital music players. Others connect (wirelessly) to your smartphone to send and receive calls and texts, as well as access other phone-based apps. Other smartwatches do fitness tracking, monitoring heart rate and physical activity. Still others record digital photos and videos. And some smartwatches, in the future, might function like full-fledged computers.

Of course, one of the most important features of a smartwatch is telling time—it is still a watch, after all. To that end, most smartwatches offer customizable displays that can look like a standard watch face, a digital watch, or a mini-computer screen. You personalize the screen however you like.

Today's smartwatches are tethered to another device you carry on your person, most often your smartphone, typically via Bluetooth technology. Tomorrow's smartwatches may be freestanding devices, with enough built-in computing power to render the connection to a smartphone unnecessary.

Samsung Galaxy Gear

The best-selling smartwatch in the pre-Apple Watch world is Samsung's Galaxy Gear 2, shown in Figure 6.1, which pairs with Samsung's Galaxy smartphones. The Gear 2 features a 1.63" display and operates either by tapping the touchscreen or speaking voice commands. The Voice Command feature enables you to perform basic functions, such as responding to texts or reading email, when your hands are full.

Figure 6.1 *Samsung's Galaxy Gear 2 smartwatch.*

The Gear 2 also functions as a standalone music player, with enough onboard storage for a medium-sized digital music collection. It also includes a built-in infrared (IR) blaster, so your watch can work as a remote control for your living room TV.

Oh, and it tells time, too. All for just $299.

Android Wear

Not to be outdone, Google has introduced its Android Wear collection of third-party smartwatches that connect to any Android device—including Samsung Galaxy phones. Google has tweaked the Android operating system (OS) to make it more wearable-friendly, and the result is a nicely integrated system that lots of companies (and consumers) can take advantage of. Current or planned Android Wear watches include the Motorola Moto 360 (shown in Figure 6.2), LG G Watch, and Samsung Galaxy Gear Live. Android Wear watches are priced in the $200 to $300 range.

Other Popular Smartwatches

Other popular smartwatches today don't have quite the feature set found in the Android Wear gear, instead tending to focus on particular functionality (such as sports or fitness tracking). These smartwatches include the COOKOO Connected Watch, Martian Watches Voice Command and Notifier, Pebble Smartwatch,

Qualcomm Toq, and Sony Smart Watch SW2. Expect to pay in the $100 to $200 range for these less-smartwatches.

Figure 6.2   *Motorola's Moto 360 Android Wear smartwatch, with a conventional dial body.*

Apple Watch

Then there's the newest player in the marketing, a little company from Cupertino called Apple. Steve Jobs' corporate progeny intends to shake up the smartwatch market much the same way the company revolutionized MP3 players with the iPod and tablets with the iPad. The new Apple Watch (what happened to the i?), expected to ship early in 2015, is shown in Figure 6.3.

Like most competing smartwatches, the Apple Watch runs a variety of popular apps (many while tethered to your nearby iPhone). You'll be able to send and receive text messages (via iMessage), Twitter and Facebook updates, and email. You also have access to Weather, Calendar, and Photos apps. The Passbook app includes the new Apple Pay wireless payment system. And there's Apple Maps to keep you on track—as well as that works.

Naturally, the Apple Watch will play back all the music in your iTunes collection. It also functions as a bit of a fitness tracker with built-in health sensors and fitness and nutrition apps.

Figure 6.3   *The many faces of the Apple Watch.*

Pricing starts at around $350 and goes up from there, depending on the style, color, and band you choose. That puts the Apple Watch smack dab at the top of the market, price-wise—even if the feature set is more grounded in the middle of the pack. Apple being Apple, of course, the Apple Watch will no doubt appeal to a ready-made base of fanboys and followers, with no compunction about lining up days in advance to pay whatever Apple demands for the latest shiny shiny.

Exercising with Fitness Trackers

Many smartwatches include fitness tracking features, but don't offer the full functionality of a dedicated fitness tracking device. Fitness and activity trackers are wearable devices that, quite simply, monitor your physical activity.

Understanding Fitness and Activity Trackers

There are several different types of fitness and activity trackers. Basic activity trackers are designed for the average person who wants to lead a healthier lifestyle. Some track your physical activity; others track your calorie intake; still others monitor your weight. The information collected is then synch'd to your smartphone or computer for further analysis and reporting.

More advanced fitness trackers are designed specifically for runners and other serious exercisers. These devices, typically in the form of a specialized smartwatch, use global position system (GPS) and other technology to record and monitor your distance, time, pace, heart rate, and other vital statistics.

Whatever type of tracker you're looking at, expect to find a variety of form factors. Some are designed to look like smartwatches; others are integrated into armbands or wristbands; still others clip on your belt or can be worn as bracelets or necklaces. Choose the one that best suits your lifestyle needs.

 Note

> Some industry experts are predicting the imminent demise of the fitness tracker market. The thinking is that the multiple-function smartwatch will replace the single-function fitness band (especially with the looming entry of Apple into the market), much as multi-function tablets replaced single-function ebook readers in that other market. Fitness tracker manufacturers argue otherwise, projecting that sports enthusiasts will want a single-purpose device rather than one that tries to be a jack-of-all-trades. While time will inevitably tell, I'd expect fitness bands to drop in price to remain attractive compared to the much more expensive smartwatches—and then see what happens from there.

Tracking the Trackers

When it comes to fitness and activity trackers, the most popular today include:

- Casio OmniSync STB1000
- Fitbit Flex, One, and Zip
- Garmin Forerunner and Vivofit
- iHealth AM3
- Jawbone UP24
- Microsoft Band
- Misfit Shine
- Polar FT2, FT4, FT7, FT40, FT60, FT80, and Loop
- Samsung Gear Fit
- Withings Pulse O2

For example, the Garmin Vivofit, shown in Figure 6.4, tracks how many steps you've taken and how far you've travelled. It also calculates calories expended and monitors your heart rate. Even more fun, it keeps track of time and lets you know when you've been inactive for too long. Nice nagging, there. It costs $129.

Figure 6.4   *The Garmin Vivofit activity tracker.*

Where the Vivofit is in the familiar band form factor, the Fitbit Zip is a less obtrusive clip-on device, as you can see in Figure 6.5. You can attach it to your belt or other piece of clothing, and it tracks steps, distance, and calories burned. It synchs to your smartphone or computer, from which you can track your progress. It's also less expensive than the Vivofit, at just $59.95, and comes in a variety of stylish colors.

Figure 6.5   *The Fitbit Zip wireless activity tracker.*

Note

Fitbit is the big dog in the fitness band market, with a 50 percent market share.

Then there's the newest and most notable entry into this collection, the Microsoft Band, shown in Figure 6.6. As the name implies, the Band is in a typical wristband configuration that displays key fitness-related data—how far you've run or walked, your heart rate, and so forth. It also includes built-in GPS tracking, so it knows where exactly you've been.

Figure 6.6   *The Microsoft Band activity tracker.*

One cool thing about the Microsoft Band it that it ties into your smartphone (iPhone, Android, or Windows Phone) and the new Microsoft Health app that helps you track and manage your health-related activities. It also displays email messages, calendar alerts, and the like, as piped from your phone. The Band runs $199, which is a tad higher than many competing devices.

 Note

Nike used to be a big player in this space, with the Nike+ FuelBand activity tracker and Nike+ SportWatch fitness tracker. But in April 2014, the company decided to get out of the tech hardware business and laid off most of the employees in its Digital Sport division.

Keeping Well with Wearable Healthcare Devices

Activity and fitness trackers are just the tip of the proverbial iceberg when it comes to health-related wearable tech. It just so happens that there are all sorts of other health-related issues that you can monitor in real time, via technology that you either wear or attach to your clothing. There's a lot of this wearable medical tech on the market today, to monitor all sorts of medical conditions.

For example, if you want to monitor your blood pressure, you can use a wireless blood pressure monitor such as the iHealth BP7 or Qardio Qardiarm. Use the Lumo Lift to monitor your posture, or the colorful SunFriend (shown in Figure 6.7) to monitor your exposure to ultraviolent (UV) light. And the HealthID Band functions as a high-tech emergency medical ID bracelet.

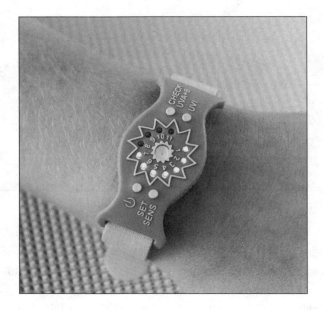

Figure 6.7 *Keep healthier in the sun with the SunFriend UV monitoring device.*

Then there's AiQ's BioMan t-shirt. This funky piece of clothing has "smart sleeves" that monitor the user's heart rate, perspiration rate, and skin temperature. That's pretty cool in and of itself, but the shirt can also be modified to measure skin mois-ture, electroencephalography (EEG), and electrocardiogram (EKG) signals. It sends its results to your smartphone or computer via Bluetooth.

In addition, OMsignal sells what the company calls "biometric smartware." We're talking sports-oriented t-shirts that, in addition to their other features (including moisture wicking fabric with strategic compression), come with little black boxes that record and stream real-time biometric data to your smartphone. As you can see in Figure 6.8, the shirt is cool looking and futuristic, and it's a great way to track all your physical performance.

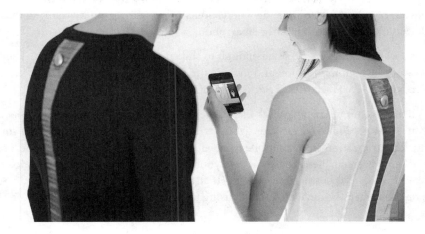

Figure 6.8   *OMsignal's smart t-shirts and accompanying iPhone app.*

 Note

These smart t-shirts aren't cheap. OMsignal sells what it calls an Up & Running Kit that includes one biosensing compression shirt, one data module, and a universal serial bus (USB) charging cable for $249. (The shirt itself, sans data module, sells for $120.)

And that's just what's available today. How about a simple patch that attaches to a home-bound patient's arm that then sends all manner of vital statistics to his or her clinic or doctor across town for remote diagnosis? Or a similar wearable that monitors a person's activities and alerts his health insurance company of what he's doing right or wrong—so the insurance company can raise his rates for bad behavior? Or contact lenses for diabetics that test the wearer's tears for glucose levels? (That latter one is actually under development by Google, believe it or not.) It's all coming. Just wait for it.

 Note

Learn more about smart medical devices in Chapter 11, "Smart Medicine: We Have the Technology..."

Monitoring Your Family with Wearable Trackers

Ever wonder where your spouse is at? Worried about your kids getting lost or distracted on their way home from school? Then check out wearable GPS-enabled devices that let you monitor the whereabouts of a given person by tracking these devices from your computer or smartphone.

Most of these devices are small enough to slide into a coat or shirt pocket. They all work in a similar fashion and include the Spy Spot TT8850 Micro Tracker, Trackimo GPS Tracker, and PocketFinder Personal GPS Locator.

When it comes to keeping track of younger children, consider the Lok8U ("locate you"—get it?) Freedom for Kids, which straps on like a wrist watch and lets you track their location from a mobile receiver. As you can see in Figure 6.9, the Trax tracker is even smaller; just slip it into your kid's pocket and then track his location from the associated smartphone app. (It also works for keeping track of your pets.) You then track the device's location using the matching smartphone app.

Figure 6.9   *Track your kids' location on your smartphone with the compact Trax tracking unit.*

And if what you really want is an instant alert in case of an emergency, there's the V.ALRT Personal Emergency Alert Device. It's a round button-like thing, not much bigger than a quarter, that you or members of your family can carry in a pocket or around your neck. One press of the button and the V.ALRT pings your smartphone and sends personalized emergency texts to three preselected contacts. The text says that help is needed and includes location information from your phone's GPS sensor.

Then there's Cuff, which offers a line of smart jewelry (bracelets and necklaces) that function as location trackers and emergency transmitters. Press a button on the Cuff and an alert (including your current location, via GPS) is sent to a list of designated recipients. Cuffs are more stylish than typical location trackers, so that's different.

Recording with Wearable Cameras

Want to know the most fun wearable tech today? It's the wearable camera, which lets you capture everything you're doing in photos and videos—from your own personal perspective.

There are two different types of wearable cameras. The first, what we'll call *action cams*, are ruggedized for use in capturing sporting activities, such as biking, skiing, or parachuting. These include the venerable GoPro (shown in Figure 6.10), Contour+2 and ContourRoam2, Garmin VIRB Elite, and Sony POV Action Cam. You can mount these cameras on a helmet or body harness, or just hold them in your hand.

Figure 6.10   *Put a camera on your head with the GoPro wearable camera.*

The next category of wearable camera we'll call the *spy cam*. These cameras are small enough to be either unobtrusive or totally unnoticeable. Most will fit in your pocket or clip on your clothing and let you record your surroundings without anyone else noticing. These wearable cameras include the Narrative Clip (which clips onto your shirt, as shown in Figure 6.11) and Autographer (designed to wear around your neck), both perfect for constant filming of what some are calling *"lifelogging."*

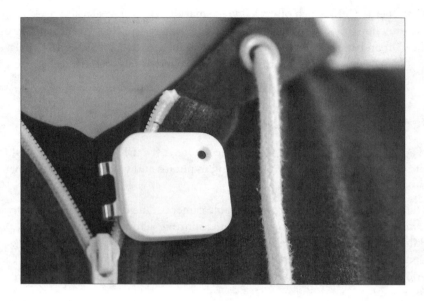

Figure 6.11 *Record everything around you, all the time, with the miniature Narrative Clip camera.*

 Note

Lifelogging (also known as lifeblogging) is the act of capturing one's entire life on camera. Lifeloggers typically use wearable cameras to record first-person video 24/7, or at least during waking hours.

Eyeing Smart Eyewear

Speaking of wearable cameras...

You've probably read about Google Glass, and how it's revolutionizing all sorts of things. Google calls it nifty futuristic wearable "smart eyewear," and I suppose it is—although it's really a lot more. (Including a wearable camera.)

Google Glass

Google Glass is essentially a miniature computer, smartphone, and digital camera all built into a set of eyeglasses, as shown in Figure 6.12. You pick your frames (some are more stylish than others), slap 'em on your face, and get ready to multitask via the voice-activated controls and miniature screen just over your right eye.

Figure 6.12 *Google Glass—trendy, fashionable, and geeky cool.*

You can use Google Glass to make phone calls, send messages, listen to music, track your fitness routine, map your location (and provide driving or biking directions), search the Web, and—oh yes—take pictures and videos of everything you see. You can also use Google Glass to make phone calls and listen to music, so you don't need to whip out your smartphone quite as often. All this makes Google Glass the ultimate wearable tech today, although it's bound to be surpassed in the future.

Not unexpectedly, Google Glass is a little on the pricey side—$1,500 to be precise. That's the price of being on the bleeding edge.

Recon Jet

Interestingly, Google Glass isn't the only glasses-like gadget designed to augment your reality. Recon Jet, shown in Figure 6.13, offers a similar heads-up display, but designed especially for athletes. It gives the wearer real-time performance metrics, such as GPS-based location, speed, distance, elevation, and more. It also connects to your smartphone for phone calls and such. And, while Recon Jet is more purpose-specific than Google Glass, it's also a lot lower-priced—just $599.

Figure 6.13   *Recon Jet—like Google Glass, but for sports activities.*

 Note

Recon Instruments also sells the Snow2, a similar heads-up display designed for skiers and built into a pair of snow goggles.

Glass Backlash

Not everyone is excited about Google Glass and other wearable eyewear. Forget the fact that it makes the wearer look like a cyborg in the making (or just super dorky), there are some very real issues aborning.

First, there's the concern that someone wearing Google Glass is taken out of the moment. Who's to say that the sweet young thing sitting across the table from you on a potentially romantic date isn't surreptitiously surfing the Web or playing solitaire via her stylish Google Glass eyewear? You know how tempting it is today to whip your smartphone to check messages or do a Google search; imagine the world if you can do all of this with a blink of the eye while pretending to pay attention to the person you're talking to. The potential for increased inattentiveness is frightening.

Then there's the privacy issue. Since you can use Google Glass and similar devices to record audio, video, and still photos of whatever it is you're looking at, who's to know that the person you're talking to isn't recording the entire conversation? Or using the opportunity to spy on the surrounding location? Or taking creepy unwanted photographs of pretty women? Or even making an unauthorized recording in a movie theater or playhouse?

For all of these reasons, and probably a few more, not everyone is looking forward to a Google Glass-enabled future. When you think about it, Glass and its ilk represent an unsettling (and unsettled) technology. It opens up new privacy and new social issues. These issues probably don't warrant legislating against the technology, although there will no doubt be some effort to do just that. It makes you wonder, though—just what rights do you give up when you start wearing a somewhat stealthy universal recording device? And how involved can you be in the real world when there's a constant gateway to the Internet literally staring you in the eye?

It's not surprising, then, to see that a lot of fans and early adapters are ditching Google Glass. Initial users (dubbed Glass "Explorers") have scaled back their day-to-day use, many selling their units on eBay for a fraction of the initial price. While developers continue to explore industry-specific apps for the hardware, development of consumer apps has dwindled to next to nothing. Several key employees in Google's Glass division have left the company (including the lead developer), and Google has quietly pushed back its planned 2014 official release of the product. It's looking less and less likely that Google Glass has a prominent place in our wearable technology future. The combination of high geek factor, low usefulness, and high price is deadly to mass adoption of any technology, and that's exactly what Google Glass is experiencing. Even Astro Teller (yes that's his name—changed from Eric), head of the Google X research labs, thinks that Glass is wildly overpriced, and that convincing people to wear smart devices on their faces is, in his words, a "tough sell." He thinks the price needs to be a quarter of the current price, tops, to garner consumer interest.

I think he's right. Google Glass at $350 would be a lot more interesting than the same product at $1,500. But even then, convincing large numbers of consumers to adopt such an intrusive piece of technology as part of their clothing isn't an easy sell. As Teller says, "It's going to be a bumpy ride."

Note

As of January 15, 2015, Google Glass is officially out of beta testing and the prototype is no longer being manufactured. While Google says Glass is "not dead," it's unlikely it will see widespread release in the near future. Instead, expect Glass to be retargeted to businesses in a limited number of key industries, and not to the general consumer market.

Wearing Other Smart Clothing

But that's not all. There's lots of other wearable tech on the market today and in the pipelines that promises to revolutionize your daily life. Just make sure you don't put any of these wearable gizmos in the washing machine with the rest of your clothes!

For example, the Razer Nabu, shown in Figure 6.14, is a wearable "smartband" designed for social media use. It delivers messages and logs your activity data, as might be expected, but also lets you interact with other Nabu wearers in your vicinity. It's supposed to learn from what you do to offer you more personalized options.

Figure 6.14   *Razer Nabu, a social smartband.*

Then there's Ring, from startup Logbar Inc., a "wearable input device" that looks like... well, a ring. This deceptive little gizmo, shown in Figure 6.15, lets you control selected home appliances and apps with a simple gesture. You can also use Ring to send messages by drawing letters in the air with your finger. And, like the best of wearable tech, no one will know you're wearing anything special.

If you have a drawer full of mismatched socks, wearable technology provides a solution in the form of Smarter Socks by the BlackSocks company. These are socks with built-in radio-frequency identification (RFID) chips. Use the associated Sock Sorter app to keep matching pairs together. (The app also keeps track of how many times your socks have been washed, in case you're interested.)

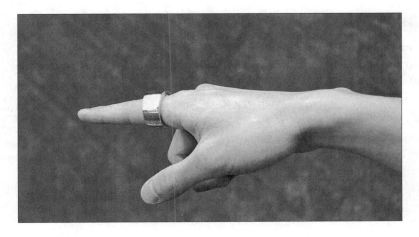

Figure 6.15   *Control your world through gestures with Ring.*

If the idea of smart socks isn't crazy enough, check out Smart Diapers from Pixie Scientific, shown in Figure 6.16. (Yes, Smart Diapers.) This wearable-for-the-infant set contains embedded sensors and a unique Quick Response (QR) code. Scan the QR code with your smartphone and the associated app analyzes the data collected, alerting you to signs of urinary tract infections, kidney problems, and the like.

Figure 6.16   *Smart Diapers for your baby.*

Hail the new frontier—of digital diapers!

Dealing with Your Personal Data

Wearable tech devices are capable of collecting a lot of personal data about the people wearing those devices. Just what happens to the information that your smart clothing knows about you?

When you consider the wearable technology of today and the foreseeable future, you note that few of these devices are self-contained. The data collected by a given device might display on that device, but most often is transferred (wirelessly, of course) to a smartphone, tablet, or computer for further analysis. An activity tracker collects and transmits data about your daily activities; smart clothing collects and transmits data about your skin temperature, heart rate, and the like; and Google Glass records and transmits images and sounds from the world around you.

 Note

A smartwatch is more of a two-way system, but still tethered to your smartphone. The wrist-mounted device receives data from your smartphone to display the contents of your message inbox and let you view photos and listen to music. Some smartwatches send activity-related data back to your phone for storage and analysis. And all smartwatches work interactively with your smartphone to manage texts, emails, calendar appointments, and the like.

In any case, your wearable tech interfaces with and shares collected data with your smartphone or computer. That's where it ends, right?

Wrong. The data that your wearable technology collects doesn't always stop at your smartphone or computer. In many cases, your personal data is beamed back to the company or service that manufactured or sponsors your wearable technology item. And this is an issue.

The Value of Data

Data has value. Data collected can be used to help improve the host product, provide key information about larger trends, and create solutions to vexing health or social issues. Data collected can also be used by advertisers, to send you targeted advertising based on your activities or condition.

Given that much of this data is biometric data about your body, it can be used by healthcare professionals to determine proper care and medication. Or it can be used by insurance companies to set your insurance rates—or even deny coverage, based on what you're doing or how healthy (or unhealthy) you are.

This raises several issues, the first (but certainly not the most important) being who profits from this data. Spoiler alert: It's not you.

When Apple or Samsung or Google or whichever entity receives data about you collected by one of its devices, it can sell that data to any number of companies. The operative word here is "sell," of course. Selling customer data is set to become a significant revenue stream for companies participating in the wearable technology market. The more data a given device collects, the more opportunities a company has to profit from that data.

You, however, don't—profit, that is. Whatever data Apple and Google collect and sell, they keep all the money generated by that data. They don't share it with you or any other customers. You have become, in effect, a tidy little revenue generator for those big companies, and you don't get paid for it. Doesn't sound cricket, does it?

It's Your Data, Isn't It?

Wait a minute, I hear you saying. This data is about *me*, about *my* body mass and temperature and respiratory rate, about how many miles I've run, and how long I sleep, and who knows what else. It's *my* data. What right does Apple or any other company have to sell it? Or, for that matter, to gather it at all?

That's a reasonable response, especially when you consider how guarded physicians and hospitals are with medical records and personal data today. It's the law, actually; healthcare professionals and institutions have to keep your medical records private. You have to give your permission for any additional use, even if it's just transferring those records from one clinic to another.

The law in question is the Health Insurance Portability and Accountability Act (HIPAA), and it's designed to protect patient data and medical records. It's a good law, if sometimes cumbersome for your doctor's office to deal with. Unfortunately, the HIPAA doesn't apply to user-generated data collected by personal wearable devices or clothing. So any data Apple, Google, and their ilk collect can be used (and sold) however they deem fit.

Unless, that is, you say they can't. Some manufacturers let you configure their devices (actually, the apps that manage those devices) to determine who can access that data and how it can be used. Some don't. And there's always the potential for privacy breaches that might expose your data, either from the device itself or with the company collecting the data. There's no clear privacy path on this as yet.

In fact, it's a safe bet that none of these companies are collecting this data with the goodwill of mankind in mind. They're in this business to make money, and they will figure out a way to generate a profit from the data their devices gather.

Managing the Data

This brings us to the development of Apple's HealthKit, a component of the company's mobile OS first introduced in September 2014 as part of iOS 8. HealthKit is essential to the operation of the new Apple Watch, but also is accessible to other device manufacturers, app developers, and the like. It's designed to make the data collected as portable as possible.

To that end, HealthKit acts as a central storage and distribution point for all manner of health-related data collected by the Apple Watch and similar devices. Already, Apple has partnered with the Mayo Clinic and Epic System (a leading electronic health records company) to manage this data and somehow make it available to physicians and other healthcare professionals.

Apple isn't the only company thinking along these lines. Google has its similar Google Fit, a health hub for user data collected from wearable Android devices. And Samsung is developing a cloud-based software platform dubbed SAMI, for analyzing and managing data collected from its wearable devices.

So there will be competing services playing in the wearable data management space. Apple, Google, and Samsung are all courting application developers with their proprietary software development kits (SDKs) for their respective services. All are courting other technology companies to tie their devices into their respective systems. All are courting healthcare data management companies to get into the healthcare end of the system. With all this activity, it's a cinch that your personal data will end up somewhere else other than your wrist or computer desktop.

Putting the Data to Use

Just how can all this collected data be used? There are lots of possibilities.

There is certainly the promise that this collected data can be of tremendous value to individual users. That's especially so when the data from multiple sources can be combined and analyzed, so that intelligent decisions can be based on that data.

Consider, for example, if your smartwatch detects a variation in your heart rate and, simultaneously, your smart shirt indicates that you're sweating and having trouble breathing. Separately, that might mean nothing. But combine the data points and a smart service might rightly determine that you're about ready to have a heart attack and then alert paramedics or your doctor of your situation.

These devices can detect more than just emergencies, of course. If your fitness tracker is monitoring your vitals on a daily basis and then sending that data to your physician, your doctor will know immediately when you need to change the dosage of your blood pressure, thyroid, or diabetes medicine. No waiting for the next doctor's appointment and the dreaded blood draw; the data flows regularly.

Even better, the large amount of data gathered can be used by experts to gain deeper insights into various medical ailments. Consider how much more we could learn about diabetes or high blood pressure with the kind of mass real-time data collected from a cross-section of the wearable tech public. Thinking in these terms, wearable technology represents a real boon for medical research.

Other uses of this data are potentially less noble, however. Think advertising here. If you start to gain a little weight or otherwise get out of shape, that data might be sold to a fitness center who then sends you targeted messages inviting you to try out their facilities. Or maybe you'll start receiving online ads for diet health drinks and personal trainers. Stuff like that.

Enter the Insurance Companies

There's one industry with particular interest in the personal data collected by all these wearable devices. We're talking the insurance industry, who aren't always looking out for your best interests.

Insurance companies are in business to make money. That's a given. The days of "sharing the risk" are long gone; they want to make a profit on every customer they have, and if they can't, they don't want that person as a customer. It's business, that's all it is.

One way to make money off a given customer is to charge that customer a rate commensurate with his likely use of the company's services. Put more succinctly, if you're healthier, they charge you lower premiums than if you're less healthy. Older individuals get charged more (because they're more likely to get sick); younger individuals get charged less. Up until the implementation of the Affordable Care Act (ACA, known on the street as Obamacare), insurance companies could even deny coverage if you had a preexisting condition; they can't lose money on you if they don't insure you at all.

This costing by risk factor is one reason why your health insurance rates fluctuate from year to year. What if, instead, your insurance company monitored your health on a daily basis and adjusted your rates just as often? If your blood pressure starts to go up, your rates do, too. (Which could cause a corresponding increase in your blood pressure when you find out about it—it's a vicious cycle.)

This is why the big insurance companies—United Health, Kaiser Foundation Group, Humana Group, Aetna, and the like—have been all over the manufacturers of wearable technology to get a seat at the table. It's not that they want to partner with the tech companies; they just want access to the data they collect. As wearable tech becomes more mainstream, that's more insured people that the insurance companies can gain more information about—and then use that information to either adjust rates or even deny coverage.

And it's not just the insurance companies. Many large employers are overly eager to gain access to this sort of data about their employees, in order to lower their health insurance costs. Imagine one of those emotionless desk jockeys in your company's human resources (HR) department, hunched over a computer dashboard monitoring the real-time biometric data of the company's employee base, determining which employees should be rewarded for their good health and which should be docked or terminated for being too fat or too lazy or eating an extra brownie for lunch. Scary but more than possible. It's the way these people work.

In fact, they're already laying the foundation for this Orwellian HR nightmare. Check out the number of fitness tracking devices from Fitbit and other companies that are being sold directly to large corporations. Then delve into your company's latest corporate wellness program and notice that fitness tracking bands are now being offered, at a special employee discount price. Then consider why your company wants you to wear a fitness band, and how it might use the data collected from it.

There are many real-world examples of this already. Take oil company BP, which gave 14,000 employees free Fitbit Zips in exchange for letting the company track their steps over a one-year period. If an employee walked more than one million steps, he or she gained wellness points that could go toward lower insurance premiums.

Now, that's positive reinforcement and that's well and good as far as it goes. But it's still using personal data to try and change individual behavior, and some feel that oversteps the bounds. What's to stop a company from using a stick instead of the carrot, as BP did? Imagine your employer taking away wellness points if employees don't meet certain activity targets. (Know that many employers already add surcharges to employees' insurance premiums if they smoke.)

It comes down to this: How much do you want your employer or your health insurance company to know about you? And how far do you trust them with the information they have?

How to Keep Your Personal Data Personal

If you're worried about the wrong parties gaining access to the private medical information collected by your wearable devices, what recourse do you have?

First, make sure you read, understand, and agree to the privacy and data sharing policies for each wearable device you own. If you don't like the way a given company shares the information gathered by its smartwatches or fitness trackers, choose a model from a company that's more conscious of your privacy.

Second, make sure you configure each device—and its accompanying apps and services—to share as little data as possible with the mother ship and associated third parties. While not all data sharing can be avoided, especially if you want to maintain full functionality of the device, turn off as much of it as you can.

Finally, if you're really concerned about your medical data being overly shared, avoid this sort of wearable technology. I know, I know, there's benefit to be had from wearing the latest smart shirt or activity band, but if you don't wear the darned thing, it can't collect any data about you. Given the importance and sensitivity of much of this information, it's a tradeoff you have to consider.

 Note

> Some users, not so willing to cede control of their personal data to third parties, are "self-tracking" with their wearable devices—that is, disabling the sending of data from their devices and monitoring their progress on their own. Some users, part of the "quantified self" movement, focus on sharing their data on their own with other individuals, discussion groups, and sharing sites. You don't have to beam your data to the mothership if you don't want to.

Where Do You Get Those Wonderful Toys?

We've talked about a lot of smart wearable devices in this chapter, most of which are readily available for purchase today. If you have the interest and the cash, here's a list of the companies we discussed:

- AiQ Smart Clothing, Inc. (BodyMan shirt), www.aiqsmartclothing.com
- Android Wear (smartwatches), www.android.com/wear/
- Apple (Apple Watch), www.apple.com/watch/
- Autographer (wearable camera), www.autographer.com
- BlackSocks (Smarter Socks), www.blacksocks.com
- Casio (OmniSync STB1000 fitness watch), www.casio-usa.com
- Contour (Contour+2, ContourRoam2 wearable cameras), www.contour.com
- COOKOO (connected watch), www.cookoowatch.com
- Cuff (smart jewelry), www.cuff.io
- Fitbit (Flex, One, Zip activity trackers), www.fitbit.com
- Garmin (Forerunner, Vivofit, VIRB Elite), www.garmin.com
- Google (Google Glass), www.google.com/glass/

- GoPro (action cameras), www.gopro.com
- HealthID (Health ID Band), www.healthid.com
- iHealth (AM3, BP7 monitors), www.ihealthlabs.com
- Jawbone (UP24 activity tracker), www.jawbone.com
- LG (G Watch), www.lg.com/gwatch/
- Logbar, Inc. (Ring controller), www.logbar.jp/ring/
- Lok8U (Freedom for Kids tracker), www.lok8u.com
- LumoBodyTech (Lumo Lift), www.lumobodytech.com
- Martian Watches (Voice Command and Notifier smartwatches), www.marianwatches.com
- Microsoft (Microsoft Band), www.microsoft.com/microsoft-band/
- Misfit (Shine monitor), www.misfitwearables.com
- Motorola (Moto 360 smartwatch), moto360.motorola.com
- Narrative (Narrative Clip camera), www.getnarrative.com
- OMsignal (biometric smartwear), www.omsignal.com
- Pebble (Pebble Smartwatch), www.getpebble.com
- Pixie Scientific (Smart Diapers), www.pixiescientific.com
- PocketFinder (Personal GPS Locator), www.pocketfinder.com
- Polar (FT2, FT4, FT7, FT40, FT60, FT80, Loop), www.polar.com
- Qualcomm (Toq smartwatch), toq.qualcomm.com
- Qardio (Qariarm monitor), www.getqardio.com
- Razer (Razer Nabu smartband), www.razerzone.com
- Recon Instruments (Recon Jet, Snow2 heads-up displays), www.reconinstruments.com
- Samsung (Galaxy Gear 2, Galaxy Gear Live, Gear Fit), www.samsung.com
- Sony (Smart Watch SW2, POV Action Cam), www.sony.com
- Spy Spot (TT8850 Micro Tracker), www.spy-spot.com
- SunFriend (UV monitor), www.sunfriend.com
- Trackimo (GPS Tracker), www.trackimo.com
- Trax (Trax tracker), www.traxfamily.com
- VSN Mobil (V.ALRT Personal Emergency Alert Device), www.vsnmobil.com/wearables/v-alrt/
- Withings (Pulse O2 activity tracker), www.withings.com

SMART CLOTHING AND YOU

You pull on your jeans, slide into your favorite tee, put your smartphone and keys in your pocket, and you're ready to boogie.

At least that's the way it is today. In the world of wearable technology, you have many more clothing-related choices to make.

Are you going for a walk or a run? Then maybe you should slip on an activity band or smartwatch with a built-in fitness tracker. If you're hardcore about your exercise, then ditch the comfy tee for a high-tech biometric-activated shirt. And don't forget to bring your smartphone, so you can download the collected data in real time.

Skip the exercise. Even if you're just heading down to your local Starbucks for a cappuccino, strap on your smartwatch. You'll use it to check your messages, listen to some tunes on the way, and maybe even pay for your drink via Apple Pay or some similar app.

If you're heading to the beach, ditch the smartwatch for a UV-monitoring wristband. If you're concerned about your blood pressure, strap on a wearable blood pressure monitor. And if you're trying to correct your posture, there's a band for that, as well.

And don't forget the kids. You want to make sure they're not wandering away or coming to harm, so clip the latest personal tracking device onto their backpacks or shirts. Heck, you want to keep track of the dog, too, so make sure he has one of those gizmos on his collar. Might as well slip one onto your spouse's belt while you're at it, you never know.

Or maybe you're really high tech and want to monitor what's going on in the world while you're out in the world. Unpack your Google Glass eyewear so you can view the latest news and stock reports (and tweets and Facebook posts) without anyone knowing. (Although they'll know. They always know.) Since you have it on anyway, might as well take a few snapshots of where you're at, or maybe immortalize your day with some livelogging video. That way you can see what you did when you get home—because you were probably too busy texting to notice it at the time.

The point is, there's a lot of wearable technology available today that you can use for a lot of different purposes. Granted, a lot of it's way too expensive to be practical, but that will change over time. You can scoff at Google Glass now when it costs $1,500 a pair but might it be a tad more appealing at a third the price? And that pricey smartwatch that looks more like a bourgeois affectation at $400 might become a must-have accessory when the price drops below $100. Even smart diapers look attractive if the price is low enough.

That's what happens with technology—it keeps marching on, just like time and OneRepublic. Tech becomes cheaper and more powerful over time. That will make it more feasible to add various smart technologies to various wearable items. And if the tech's there, we'll find a way to use it.

Should you head out to buy some wearable tech today—a smartwatch, maybe, or a biometric t-shirt? It depends. Some wearable tech already delivers practical value at a reasonable price; I'm thinking fitness and activity trackers, especially. If you're big into exercise, or just want to watch your health, these gizmos help you do it and don't cost an arm and a leg. Their time has definitely come.

Smartwatches are a different beast. I look at these $300 devices and wonder what they do that my smartphone doesn't already. Plus, I already have a smartphone, so why spend the $300 for a duplicative device? And who wears watches these days, anyway? If I want to know that time it is, my smartphone will do the job.

So I'm less hot on smartwatches, but that may be a personal thing. Or a price thing. Or a functionality thing. Given the focus on health-related features in upcoming smartwatches, that might be something unique and uniquely useful. At the right price, it might change my mind.

As to all the more futuristic wearable tech—smart shirts and Google Glass and all that—it's okay to take a more wait-and-see attitude. They're more proof of purpose now, and certainly not in any way affordable. Doesn't mean they won't be affordable a few years from now, however, and it's likely somebody will figure out some cool things to do with the tech. Until then, let somebody else do the guinea pig thing with smart eyewear, smart muscle shirts, and smart diapers; you can jump in when they start selling them at Target and Kohl's.

7

Smart Shopping: They Know What You Want Before You Know You Want It

The Internet of Things is changing the way you shop. Location-aware devices can sense your presence in a store and beam you context-sensitive offers and instructions. Soon you'll be able to check out without opening your wallet, thanks to RFID and NFC technologies. That's if you even have to go to the store, of course; your smart kitchen may compile a shopping list automatically and beam it to your local store, which will deliver your groceries via drone.

If that sounds farfetched, think again. The IoT is already easing into the supply chain, helping to make warehousing and inventory more efficient. Soon it will make shopping more efficient, helping to guide you to the best bargains and the products you need—or at least the ones the retailer wants to sell you the most.

Eliminating the Need to Shop

We've previously discussed the concept of the smart kitchen. Your smart refrigerator knows exactly what you have on hand and what you need to replenish. When you send it a recipe (probably beamed from your smart TV while you're watching the Cooking Channel), the fridge figures out what ingredients you need to buy and automatically compiles a shopping list.

Today, you'd have to physically take that shopping list to the grocery store and fill up that shopping cart by hand. Not too far in the future, however, it's easy to envision that shopping list being transmitted electronically to your grocery store of choice, which then fills your order and either has it available for curbside pickup or delivers it to your home. (You don't need drones for delivery in this scenario; a minimum-wage teenaged driver will do the job just fine.)

In this scenario, you no longer have to shop for groceries. You order exactly what you need and have it delivered without ever setting foot in the store. This is more efficient for you, in a couple of ways. First, you don't have to spend the time driving to the store and pushing that wobbly cart down the aisles. Second, you only buy what you need; you're not influenced by what's displayed on the endcaps or pushed at you by the friendly demonstration lady. And let's not even discuss the hazards of grocery shopping on an empty stomach...

The long and short of it is, if you can automate your grocery shopping, you'll spend less time and money doing it. That's a good thing.

While grocery shopping is yet to be significantly influenced by the wide world of the Internet, other types of shopping have been in flux for some time now. The rise of online shopping (and mega retailers such as Amazon.com) have changed the way most of us shop for clothing, toys, books, and all manner of goods. We see something we want, we tap a few keys on the old computer (or buttons on the smartphone), and voila! The order is placed and readied for delivery within two or three days.

One of the great things about online shopping is the near-immediate gratification of buying something while you're thinking of it. You don't to wait until you have a free evening to drive to the store and fight the maddening crowds; you think it, you buy it; it's that easy. (Delivery, of course, dampens that immediate gratification thing, but it's getting faster.)

The other thing about online shopping is that you're prone to making fewer impulse purchases. You search for what you want, you find the best price, you click the Buy button. If you go to a brick-and-mortar retailer for the same item, no doubt you'll find something else you want, too, and end up with a shopping cart full of stuff. Shop online, spend less money. (At least, that's the way it's supposed to work.)

Now imagine that you can automate at least some of your non-grocery shopping. This is best done with commodity items, such as toilet paper, razor blades, laundry detergent, and the like—items you can buy from the grocery store (or big box retailer) but that aren't stored in your refrigerator. Since your smart fridge can't keep track of these items, you instead need other smart storage. In essence, you have sensors in your closet, cabinet, or bathroom vanity that track specific items, probably via RFID or some similar technology. (Cruder sensors could simply employ video cameras to "read" your current inventory.) In any case, your smart storage knows what you're supposed to have on hand, and then automatically reorders items as you use them. Orders could be sent to your local grocery store or ordered from Amazon.com or a similar online merchant for to-your-door delivery.

I'm not sure that there's much more shopping that could be automated in this fashion. No sensor is going to know when you want to buy a new book or a pair of new shoes or even new toys for your kids. These items need to be shopped for manually—either in person or online. So there will always be shopping to do.

Changing the Retail Environment

Even though a lot of physical shopping has moved online, and even more will soon be automated, the world of retail is not going away. People still like to shop and have to shop for all manner of things. The exact nature of that retail experience, however, is destined to change with the advent of the IoT.

For shoppers, the promise of the IoT is a more personalized shopping experience. For retailers, the IoT promises to better target the needs of individual shoppers, and hopefully persuade them to make a purchase.

These potential benefits are two sides of the same coin, enabled by technology that identifies individual shoppers when they enter a store, calls up their personal data and shopping history, and then provides pointers and promotions that drive them to specific areas or products within the store. The smart store knows its customers and what they like, and then tailors the shopping experience accordingly.

Here's how it works.

When you walk into a store, the store knows you're there because it's tracking your smartphone.

Once the store knows you're inside, it looks up your personal data and purchasing history in its own database. If you live alone in an apartment, the store knows that. If you live with your spouse and three kids in a sprawling suburban house, the store knows that. The store also knows that you purchased a new winter coat last month, or that you purchase a particular brand of shampoo every 30 days.

The store then uses this information to tailor promotions specific to you. If you're a single apartment dweller, the store might devise a special on frozen pizza. If you live with your family in a big house, and it's winter time, the store might devise a special on snow shovels or sidewalk salt. If you purchase the same shampoo every month, it might devise a special on that shampoo.

You get notified of these specials via your smartphone, of course. They might come in the form of a text message with a digital coupon attached, or even an alert to the store's smartphone app, such as the one in Figure 7.1.

Figure 7.1 *An in-store digital coupon, beamed to a customer's smartphone.*

The store might also use this information to direct you to specific areas or merchandise. You might receive a text message or an app alert suggesting that you head over to Aisle 12 to see something special, or to the second floor sporting goods department for a personalized demonstration.

And if you put an item in your shopping cart, the smart store will know, thanks to each item's RFID tag. The store can then recommend related items or accessories you also might want to purchase. For example, if you place a dress shirt in the cart, you might get a text message or smartphone alert directing you to the store's selection of ties, with accompanying discount coupon. Or, if you're in the grocery store, the smart cart can send you recipes based on the items you've selected, along with coupons for additional items to complete those recipes.

There are other ways to use this type of personalized customer information. Intelligent product displays can use smartphone IDs or even video cameras with facial recognition technology to detect who's around them. The displays can then deliver content customized to those individual shoppers. So if you're standing in front of a display for toothpaste, you'd see different ads and offers than would your next-door neighbor.

This information also provides insight useful for longer-term customer management. Using the same tracking technologies, a smart store can effectively monitor you as you move through the store, tracking your path and the products you look at. As the store learns how you and other customers shop, this data can be used to optimize store layout for different types of customers.

Smart Store Tech

Several different technologies—all available today—will be utilized in the smart store of tomorrow.

First, individual items in the store must have RFID tags, like the one shown in Figure 7.2. RFID, short for *radio frequency identification*, is a technology that uses special tags encoded with product-specific information. Information on the tag is transmitted via short-range radio signal to RFID reader devices. Today, RFID readers are located at a store's checkout (and often in the store's warehouse—which we'll discuss in a moment). In the future, RFID readers might be attached to shopping carts or just scattered throughout the store at key points.

Figure 7.2   *An RFID tag, like the type attached to most in-store merchandise.*

Next, the store must have some way of both identifying individual customers and beaming information to them. This will be accomplished by an indoor positioning system—essentially, a combination transmitter/receiver that picks up signals from customers' smartphones.

One such system currently in use is Apple's iBeacon, shown in Figure 7.3. The iBeacon system communicates with customers' smartphones using Bluetooth technology. Multiple iBeacons—small enough to mount unobtrusively on shelves or walls—are placed throughout the store. They then work together to triangulate a phone's location, so the store knows exactly where each customer is at any point in time. The iBeacon system can then beam notifications about nearby items on sale to the customer's phone.

Figure 7.3   *The iBeacon system works within a store to locate and communicate with customers' smartphones.*

Finally, the store's checkout system must be equipped to receive NFC signals from customers' smartphones and smart watches. NFC, short for *near-field communication*, is a wireless technology that uses an electric or magnetic field (but *not* radio waves) to transmit data over a short distance. In an NFC payment system, payment data stored on a customer's phone is transmitted via NFC to the store's payment terminal. No manual credit card swiping is necessary; just hold the phone up to the terminal and NFC does the rest, wirelessly.

Making It Easier to Pay

It makes sense that retailers would want to make it easier for you to shop for items in their stores. It also makes sense that they'd want to make it easier for you to give them their money, in the form of a more efficient checkout experience.

Nothing burns my butt more than, after spending a half hour shopping, being forced to spend an equal amount of time waiting in the checkout line. Don't make it so hard for me to pay—after all, I might just change my mind and abandon my shopping cart, then and there.

Enhanced checkout is the goal, even to the point of eliminating traditional point of sale (POS) systems. The more seamless the checkout process, the better. The ideal is to walk into a store, grab what you want, and then leave, with the whole payment thing being handled in the background. Indeed, some have said that the ultimate checkout experience will feel just like stealing; there won't be any interaction with a cashier to slow your progress.

This isn't as farfetched as it seems. Thanks to RFID, NFC, Bluetooth, Wi-Fi, and other wireless technologies, every individual item in the store can be tracked. You, of course, can also be tracked via your smartphone. So if all the sensors get together and combine their information, the smart store will know that you've picked up a specific pair of shoes and are walking out the door with them. Assuming you're not a shoplifter, the store can then bill your account (via your preferred payment method) for those shoes. No waiting in line, no fumbling for credit cards, no dealing with a surly cashier.

Even without this sort of invisible background check out, the checkout process can be streamlined via the use of NFC or other wireless payment from your smartphone. This is, after all, what Apple Pay is all about, and everybody knows Apple Pay is the Next Big Thing.

With Apple Pay, you connect all your credit cards and such to a single Apple Pay account. This data is stored in a digital wallet on your smartphone or smart watch. When you're ready to pay, you open the Apple Pay app, select the credit card you want to use, and then use NFC technology (or just scan a Quick Response [QR] code from your smartphone screen) to transfer the card data from your phone to the store's terminal, as shown in Figure 7.4. It's not seamless, but it is quicker than swiping a credit card.

Apple Pay isn't the only digital payment system out there. Google Pay and Square have been trying to muscle into this space for some time now, with little success. (It's possible that Apple will be more successful, if only because of the partnerships it's forged with credit card companies and major retailers.)

Figure 7.4   *Paying digitally via Apple Pay.*

Then there's CurrentC, which is supported by a consortium of major retailers, including Target, Walmart, Kohl's, Best Buy, Sears, CVS, and 7-Eleven. Unlike Apple Pay, which uses NFC technology, CurrentC is based on QR codes. When you're ready to pay, the CurrentC app displays a QR code onscreen, as shown in Figure 7.5, which is read by a QR reader built into the retailer's payment terminal. It may seem like old tech, but QR codes work and are fairly secure—plus, there are all those big retailers buying in.

Figure 7.5   *The QR payment screen in the CurrentC app.*

Note

A QR (Quick Response) code is a type of barcode that contains information about the item to which it is attached.

Whatever technology or system ends up ahead, the goal is to speed up the in-store payment process. If that payment process can be further integrated into your shopping experience, even better.

Deliveries by Drone

Let's return for a moment to those customers who don't shop in bricks-and-mortar stores. When you place an order online (either manually or via smart automation), how does that item get delivered to you?

Today, an item you order from Amazon or a similar online retailer is likely to be delivered by UPS, FedEx, or the U.S. Postal Service. Even with the best of systems, you're talking a two-day or more wait between your ordering and having the package show up on your doorstep.

But what if the delivery process itself could be automated as part of the IoT? That's the promise of delivery drones, such as those proposed by Amazon, Google, and even Domino's Pizza. There's a lot going on in this space, which we'll discuss in Chapter 9, "Smart Aircraft: Invasion of the Drones." Turn there to learn more.

Managing Inventory Smarter

Placing RFID tags on every item in a store is the very definition of the Internet of Things. Okay, so each item doesn't have its own IP address, but it does have its own unique identity. And once an item is read into the system, all sorts of good things can happen.

Many retailers today are using RFID technology to better manage their inventory, not just in the store but throughout the entire distribution chain. A retailer can track an individual product from when it arrives in the warehouse through distribution to an individual store to the item leaving the store after being purchased. The retailer knows exactly where each piece of merchandise is at any given time.

With the right inventory management systems, the retailer can take the information gathered via RFID tracking and know how well that product or product category is performing, when to order more inventory, and even when to restock the shelves before they are completely empty. It's piece-by-piece inventory management, no matter how big or how small the item.

Today, this is all done with RFID tags. In the future, products may have smart sensors attached that transmit the same or even more data via Wi-Fi or other more advanced IP technology. That's the promise of the IoT, in any case.

What About Your Data?

One key to the success of the IoT at retail is the collection of customer data, and then the transmittal and use of that data within the store. This brings up the expected privacy concerns: How will retailers use your personal data, and how safe will it be?

Those are good questions, but there aren't necessarily good answers. Retailers do not have a great history in terms of safeguarding customer data.

First, let's address how customer data can be used. While most retailers have detailed privacy policies (that customers never read), it still boils down to if they collect it, they can use it—in whatever ways they deem fit. That could be for internal use, as we've been discussing, but the data could also be sold to other parties for their use. This is how you get on so many different mailing lists today.

In terms of in-store use, it sounds cool to have the store send you a coupon based on your past purchases, what you have in your shopping cart, or what aisle you're in. At some point, however, all this attention gets a little creepy in a stalkerish way. Do you really want to be tracked down the store aisles? We're talking unwanted surveillance here, at least in the eyes of many consumers.

Now we come to the security of your personal data. Most retailers talk a good game about how secure their systems are, but history proves otherwise. Within the past 12 months, we've seen major customer data breaches from Target and Home Depot that affected hundreds of millions of consumers. They may say your data is safe, but it's obviously not.

So the more data a retailer collects about you—and the more that data gets bandied about—the more likely it is that it will be grabbed by people you don't want to have it. Systems will get safer, but hackers will also get smarter. That's the way the cycle goes.

What can you do about all this? If you don't want to be digitally stalked while you're shopping, you can always turn off your mobile phone. (Of course, then you won't be able to get texts or calls from friends—or use your phone to do comparison shopping with other retailers.) If you don't want your personal data exposed to nasty intruder types, that's a bigger issue—and one we'll discuss in Chapter 15, "Smart Problems: Big Brother Is Watching You." Turn there to learn more.

SMART SHOPPING AND YOU

If you're a retailer, all this smart shopping stuff is pretty exciting—and challenging. It's going to require a significant investment in new technology and systems, but the rewards (in terms of increased sales per customer) are potentially huge. Or so we're led to believe.

If you're a consumer, you probably look on all this with mixed emotions. On one hand, if technology can help you save money and spend less time shopping, that's a good thing. On the other hand, if all this wonderful technology is used solely to bombard you with more unwanted ads and promotions, then that's something else.

You do have some control over how you interface with emerging smart shopping technology. If you don't find thrilling the advent of digital payment systems, simply don't install those apps on your phone. No one's forcing you to use Apple Pay, CurrentC, or any of those systems. Retailers will likely hold on to their old-fashioned credit card terminals for some time. And you can always pay with cash, the least traceable payment method out there.

As far as being tracked within the store, look for ways to opt out of participation. It's likely that retailers will recognize that many customers don't want such an immersive experience and will offer ways to opt out of all the tracking and ad-sending, much the same way they do with email and mobile marketing today. If a retailer insists on forcing its digital promos on you, then shop elsewhere. Retailers who go too far, intrusion-wise, will quickly learn from their mistakes when their sales drop.

All that said, there's something a little exciting about retailers going to all this expense just to have a more personal relationship with you. (And their millions of other customers, of course.) Sometimes it's nice to have somebody pay attention to you, even if it is just some computer in your local big box store. It's just possible that the newfound attention will be accompanied by some money-saving digital coupons. And who doesn't like to save a little money?

8

Smart Cars: Connecting on the Road

The Internet of Things (IoT) holds a lot of promise for the automobile industry, and for drivers everywhere. Today's so-called connected cars offer streaming music via the Internet, traffic and weather reports, and even global positioning system (GPS) mapping and directions. But what about a smart car that monitors traffic conditions and automatically reroutes you when necessary? Or one that diagnoses—and even repairs—its own problems if they develop? Or the ultimate smart automobile, a self-driving car that's so advanced you don't have to do much of anything to get from point A to point B other than open the door and strap yourself in.

It's obvious that the Internet of Things is going to change the way we get around, and big time. Let's find out more.

Smart Cars Today—and Tomorrow

Automobiles have been getting progressively smarter over the years. Today's cars contain dozens of computers or computer-like devices, all working together to make sure your car runs as well as it's designed to. These electronic control units control all sorts of in-car functions, including braking and cruise control, and heating/cooling and entertainment systems.

This smarting up of the family wheels is something that's been taking place over the past three or more decades. Go back to 1977, as an example, and you find that the Oldsmobile Toronado had a single computer unit that controlled spark plug timing. In the early 1980s, additional computers were introduced to improve emissions systems, and in the late 1980s, electronic throttle control (ETC) replaced cables and mechanical linkages. And on it goes.

With everything that's computerized in today's cars, it's not far-fetched to consider that the set of wheels sitting in your garage is the smartest smart device in your entire home.

 Note

> Don't confuse the concept of the smart car, with a lowercase "s," with the literal Smart car, a cute little (and I do mean little) electric car sold by the Smart Automobile company, a division of Daimler-Benz (the Mercedes folks).

Smart Functionality

Let's talk first about how computers have taken over much of the basic functionality of the average automobile. Many key mechanical systems are now controlled by microprocessors, with the goal of more efficient and safer operation.

Some mechanics look at today's automated automobiles and see a computer on wheels. Actually, it's more like 30 or more computers on wheels (close to 100 in luxury cars), all acting in concert to keep your car working in tip-top condition.

What operations are computer-controlled in a typical car? Here's just a short list:

- Air bag systems
- Anti-lock brakes
- Automatic seat positioning
- Automatic transmission
- Climate control system

- Cruise control
- Entertainment system
- Idle speed
- Keyless entry system
- Security system

These operations are supplemented by a variety of electronic sensors that feed real-time information back to the computer brains. We're talking air pressure sensors, air temperature sensors, engine temperature sensors, knock sensors oxygen sensors, throttle position sensors, and more. Data from these sensors is used to control spark plugs, fuel injectors, and other key components.

The most advanced computer in cars today is the *engine control unit*, or ECU, such as the one shown in Figure 8.1. The ECU monitors outputs from dozens of different sensors to control the engine's operation, emissions, and fuel economy. Today's ECUs are relatively low power, typically using 32-bit architecture, 40MHz CPU, and 1MB or so of RAM. That's a lot less robust than the 64-bit CPU in your personal computer, but it's all that's needed to run the simple code that controls today's automobiles.

Figure 8.1 *The engine control unit from a BMW.*

Smart Diagnostics

The data collected by a car's electronic sensors is fed to a central communications module, which stores the data for appropriate future use. If sensors detect something is amiss, the communications module alerts the driver, typically via the much-despised Check Engine light. When the driver takes the car in for service, the service technician can then extract a diagnostic code from the car computer to find out what's wrong and then fix it.

As advanced as all this sounds, it really isn't. The onboard computer doesn't actually diagnose anything (and certainly doesn't fix anything itself); it merely stores the error codes reported by other sensors in the car. The service technician doesn't get a screen full of Star Trek-style blueprints and instructions, but rather a list of simple numeric codes—which he then has to look up to see what they mean. It's better than getting no diagnostic help at all, but there's a way to go before these diagnostics can be considered in any way intelligent.

Intelligent or not, many old-line mechanics (and most do-it-yourselfers) decry this rampant computerization of what used to be simple mechanical machines. Long gone are the days where a weekend mechanic could break out the toolbox, get his hands dirty, and pound the typical ailment into submission. Instead, today's service techs need their own banks of computers to diagnose and fix problems related to the many computer-controlled operations in the current batch of state-of-the-art driving machines. You really can't fix your car yourself anymore; you have to take it into an accredited service center.

In the future, however, diagnostics will get easier because the car's electronics will be even smarter. Instead of generating indecipherable error codes, your car's smarter computer will display plain-English explanations of what's wrong. Your car's dumb warning lights will get smarter, too. Instead of displaying a single Low Tire Pressure light (and leaving you to figure out which of the four tires needs air—a time-consuming job), the in-dash display will tell you that the driver's side front tire is five pounds low. That's a simple thing that will make your life immeasurably easier.

Same thing with everybody's least favorite dashboard alert, the Check Engine light. Nobody knows what it means when this light goes on; it could be nothing, it could be something extremely serious—you just don't know until you take it into the dealer. Smart systems would do away with this overly generic warning and instead use the in-dash display to tell you precisely what is wrong—and what you should do about it.

Ideally, smart diagnostics systems will identify potential problems before they become real problems—oil levels getting low, belts wearing out, that sort of thing. And then the computer won't just tell you about the problem, it will use available wireless technologies (Wi-Fi if you're near a hotspot, cellular if you're not) to notify the repair center of the problem. The repair center will then order the necessary parts and contact you to schedule an appointment to perform the repairs.

Less guesswork, more automation. That's smart diagnostics in the future.

Smarter Driving

Smart driving systems are slowly but surely finding their way into the average family auto. We're not talking self-driving cars (although we will in just a few pages), but rather dedicated systems that help drivers perform difficult or sometimes dangerous maneuvers and operations.

The most common smart driving system is the now ubiquitous cruise control. Basic cruise control systems have been around for decades, limiting speed to a preset level. Newer adaptive cruise control systems, however, take this one step further by monitoring the distance to the car in front of you and then adapting the speed as necessary. Used to be you had to slam on the brakes or toggle off the cruise control when you ran into heavy traffic; adaptive cruise control systems do this for you and help you keep in step with the flow of traffic.

On the safety side of things, many cars today come with lane assist systems that monitor your position in your lane and keep you from drifting into the adjacent lane. You may get a warning alert if you start to wander over or, in some luxury models, the car itself may take corrective action.

Parking assist systems are also becoming more common. These systems help with the challenging task (for many) of parallel parking. For example, Volkswagen's aptly-named Park Assist system automatically detects the nearest empty parking space, measures the space, notes the current position of your car, and then carries out the optimum steering movements to put your car in its place. (You still have to operate the accelerator and the brake during the process.) The process is illustrated in Figure 8.2.

And there's more, especially in higher-end vehicles:

- The Infiniti Q50 utilizes video cameras, radar, and other technology for its lane-keeping, cruise control, and collision avoidance systems.
- The Mercedes S-Class includes systems for autonomous steering, parking, accident avoidance, and driver fatigue detection.
- Some BMWs can read speed limit signs and automatically keep your car under (or at) the proper speed.
- Tesla Motors' AutoPilot mode controls the car's steering, braking, and speed; Teslas also include auto parking.

- Volvo's 2014 models include adaptive cruise control with a steering assist feature that enables your vehicle to follow the car ahead, even if it changes lanes.

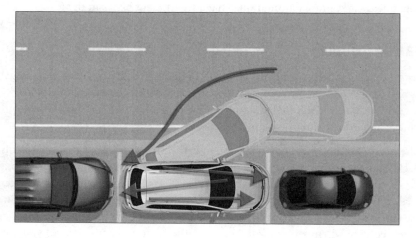

Figure 8.2   *How Volkswagen's Park Assist system automatically parks a car.*

And there's more to come—which you'll learn in the "Cars That Drive Themselves" section later in this chapter.

 Note

> Part and parcel with today's smart driving systems are an assortment of in-car cameras, sensors, and warning systems designed to provide a safer driving experience. Backup cameras are becoming standard equipment on even low-end vehicles. Side-mounted "blind spot" cameras are incorporated into lane change avoidance systems. And parking sensors let you know if you're getting too close to the cars around you.

Smart Communications

All the sensors and systems in your car don't operate in a vacuum. These devices communicate and interact with one another—and with the future—with similar devices in other cars.

Today's in-car systems typically operate via wired (not wireless) connections. The dominate communications standard is called CAN, for *controller-area networking*. This standard enables communication speeds up to 500Kbps, which is considerably slower than your home network but fast enough for the simple data transfers between your car's devices.

In the future, cars will get smarter by communicating with each other. Instead of relying on proximity sensors or radar to identify a nearby vehicle, your car will instead receive a radio signal from that other car. In essence, your car's computer brain will be communicating with that car's computer brain. Your car will know what that other car is going to do—where it's going, by what route, and how fast. By sharing this information in real time, your car can then calculate the appropriate direction and speed to not only avoid contact with that car, but also determine the best route to travel together or apart.

Car-to-car communication systems are part and parcel of the self-driving car, and of creating a safer driving experience. According to the National Highway Traffic Safety Administration (NHTSA), this sort of car-to-car communication could help to prevent more than 80 percent of all traffic accidents. That would be technology well used.

Smart Entertainment

One of the most common applications of technology in the car comes in the form of intelligent entertainment systems. It's a matter of making it easier to listen to your favorite tunes while driving.

Today's state-of-the-art in-car entertainment system not only includes an AM/FM/satellite radio (and sometimes a CD player), but also a USB port to which you can connect your smartphone or USB memory stick. The smarter of these systems can control your iPhone (though typically not an Android phone) from the dashboard display and even show album artwork for the current selection.

Even more convenient, many in-car entertainment systems let you connect your smartphone without a cable, via Bluetooth wireless technology. When I start up my Honda CR-V, it automatically recognizes my Samsung smartphone and starts playing the next tune in the phone's digital queue. I don't have to do anything other than select the auxiliary input—or, if I don't like a song, hit the Next button. All the connections happen in the background, and that's great.

Some cars even come with built-in apps for certain streaming music services, so that you can initiate playback direct from your dashboard (with your smartphone connected, of course). Pandora apps are the most common, although some cars have apps for Spotify, iHeartRadio, and other services, as well.

Apple's CarPlay technology looks to make in-car music playback even easier. CarPlay utilizes Apple's Siri voice recognition technology to let you control your entertainment system with a few well-chosen words—which is safer than turning a dial or tapping a touchscreen. (Figure 8.3 shows the CarPlay system at work.) The CarPlay system also enables control of mapping/directions (via the much-maligned Apple Maps app), text messages, phone calls, and the like.

Figure 8.3 *Apple's CarPlay system.*

Apple is working with most major automobile manufacturers to integrate CarPlay into their in-car entertainment systems. Expect to see CarPlay technology in cars from Audi, BMW, Chevrolet, Chrysler, Dodge, Ford, Honda, Hyundai, Mazda, Mercedes, Nissan, Suzuki, Toyota, Volvo, and more within the next few years—with select 2015 models already lined up.

 Note

Voice command isn't limited to Apple's CarPlay system. Other manufacturers offer similar voice control for making phone calls and answering text and email messages. (Some cars will even read your text messages for you, so you don't have to text and drive—which is becoming increasingly illegal in many jurisdictions.)

Apple isn't alone in wanting to control your car's entertainment system. Google is developing Android Auto, which controls your entertainment system, maps and directions (via Google Maps), and such from your Android phone. (Figure 8.4 shows an Android Auto screen with map and music information.) Like CarPlay, Android Auto can be controlled from your phone, your car's in-dash touchscreen display, or voice commands. Google has partnerships with many of the same automakers as does Apple, so it remains to be seen which of these two competing systems will win the battle for in-car control.

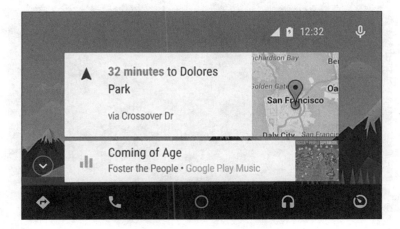

Figure 8.4 *Google's Android Auto system.*

Smart Climate Control

In-car climate control systems are also getting smarter. The day of the single-knob heater/air conditioner (turn it to the right for more heat, to the left for more air) is long gone. Today's typical system lets you set the precise temperature you want, and then turns itself on and off to keep it there. You also get multiple-zone control, so that the driver and the passenger (and maybe the kids in the back seat) can set their own individual temperature zones. It's all computer-controlled and works relatively well. (Figure 8.5 shows a typical dual-zone climate control system, part of Ford's MyFord Touch communications/control system.)

Climate control will only get smarter in the years ahead. Expect your car to recognize who's driving (by the particular key fob you carry) and automatically adjust the temperature to your preference. It'll also adjust your seat and dial into your favorite radio (or Internet radio) station at the desired volume level. It's all about personalizing the driving experience, based on what you individually like or don't like.

Hacking a Smart Car

All these smart electronics seem well and good, and no doubt we'll eventually wonder how we ever drove without them. But they also represent an opportunity for malicious hackers to take control of your car. And that's something to worry about.

Figure 8.5 *Dual-zone climate control in the MyFord Touch system.*

Here's the scenario. You're driving in your smart car down the highway at 70 or so miles per hour. With no warning, the steering wheel turns hard to the left and you crash into the car in the next lane. Or maybe your car just slows to a stop. Or shuts down altogether.

The cause of this apparent malfunction? Not your car, but rather someone hacking into your car's computer systems. It's not that far-fetched, especially when smart cars start communicating with the outside world via wireless or mobile networks.

In fact, the likelihood of malicious intrusion is higher than most car manufacturers would like to admit. That's because today's car control systems are rather primitive and, unlike your home computer or smartphone, have little to no built-in security. In other words, a car is relatively easy to hack if someone sets their mind to it.

That said, auto manufacturers are aware of the threat and are working to improve their security systems. In fact, Continental, one of the biggest auto parts suppliers, is partnering with IBM and Cisco to add firewalls to their electronic devices. Ford and Toyota are also developing firewalls to protect their cars, with the latter embedding security chips in its in-car computers.

It's likely, then, that smart cars will also become more secure cars in the years to come. But the security issue exists and could become problematic for some manufacturers.

Note

Car hacking doesn't have to be malicious. Today, third-party On-Board Diagnostic (OBD) systems let you hack into your car's computer and run your own diagnostics. With the help of these devices, you can see how other drivers are using your car, keep performance at peak efficiency, and even find out just what's behind a pernicious Check Engine Light situation. For more information, go to the OBD-II website at www.obdii.com.

Cars That Drive Themselves

Now we come to the part of this chapter everybody's been waiting for, the section about self-driving cars. Yes, they're coming—and driving themselves!

How Self-Driving Cars Work

A self-driving car is simply a car that is capable of driving itself, with little or no input from a human driver. It uses sensors, computers, and other smart technologies to sense where it is in relation to other vehicles (and the road), and navigates according to preset coordinates.

A number of different technologies need to be employed for a car to drive completely autonomously. These include:

- 360-degree cameras, to view all sides of the car
- Adaptive cruise control, to regulate speed in traffic
- Emergency brake and steering assistance, to avoid collisions
- GPS, to determine precise location and navigate routes
- Radar and Light Detection and Ranging (LIDAR), to sense the distance between your car and other cars or objects
- Stereo cameras, to outline and identify pedestrians and bicyclists as different objects than other cars

 Note

LIDAR is a technology that measures distance by illuminating an object with a laser and then analyzing the reflected light. Radar does much the same thing, but with radio waves.

Of course, a self-driving car must also have a fairly robust computer in charge of the whole shebang. The in-car computer must tie together all these sensors and systems, and determine what actions to take in any given situation. In addition, the in-car computer will also handle all the necessary routing functions, based on GPS and known maps.

I'm particularly fascinated with emergency brake and steering assistance systems, which enable the car to undertake evasive maneuvers by steering itself out of and back into a lane when it detects an obstacle. Figure 8.6 shows one such system, from parts supplier Continental: the Emergency Steer Assist system. As you can see in the figure, the system is alerted to a slower car or obstacle in enough time for the system to take over the breaking and steering systems, slowing the car down and steering it into the next lane—faster and more accurately than most human drivers can respond.

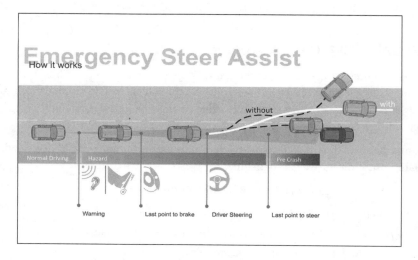

Figure 8.6 *How Continental's Emergency Steer Assist system works.*

By combining this type of collision avoidance system with adaptive cruise control, a car can drive down the road without hitting anything in its way. The exact route is specified via GPS mapping and navigation, of course. Program in your destination (or just say it; might as well add voice control to the mix) and the car's computer calculates the best route there. The automatic systems then take over and drive the route.

What's Coming

Many of the automatic systems needed for a self-driving car are in existence today. Not surprisingly, most automobile manufacturers have big plans for their own autonomous vehicles. Here's what some of the major car companies have in store over the next couple of years:

- Audi plans to market vehicles that can steer, accelerate, and brake, all by themselves.
- Cadillac plans to introduce select models with autonomous lane keeping, speed control, and brake control.
- Mercedes plans to introduce Autobahn Pilot (in the United States, Highway Pilot), a system that enables hands-free highway driving with autonomous overtaking of other vehicles.
- Toyota plans to roll out what it calls near-autonomous vehicles with Automated Highway Assist, Lane Trace Control, and Cooperative/Adaptive Cruise Control.

Looking just a little bit further into the future, Continental, the big auto parts supplier, is working on various autonomous driving technologies in its Advanced Driver Assistance Systems unit. The company plans to have autonomous assistance available for limited freeway driving and for use in construction areas by 2015. By 2017, the company expects to add autonomous low-speed city driving, with technology for driving on two-lane highways and country roads by the end of the decade. Fully autonomous driving is expected to be available by 2025, with premium and luxury cars getting the technology first.

By the way, there actually is one self-driving vehicle commercially available today. It's not a car, however. As you can see in Figure 8.7, the Navia, from Induct Technology, is a cross between a golf cart and an open-air shuttle. It's designed for use in pedestrian zones and travels at a breathtaking 12.5 mph. But it is completely autonomous, which gives us a glimpse of what's coming in more traditional passenger vehicles.

Levels of Automation

It's unlikely, however, that the self-driving vehicle will arrive in one fell swoop. The robot takeover of our roads is probably going to happen in several evolutionary steps. In fact, the NHTSA has established an official classification system that neatly outlines different levels of automation:

- **Level 0**—The driver completely controls the vehicle at all times. (This is what we have today, more or less.)
- **Level 1**—Individual vehicle controls, such as automatic braking or electronic stability controls, are automated. We're just now entering this era.

- **Level 2**—At least two controls can be automated in unison, such as adaptive cruise control in combination with lane keeping.
- **Level 3**—The driver can turn over control of all safety-critical functions in certain conditions. At this stage of the game, the car senses when it's in over its head and lets the driver know he needs to resume control.
- **Level 4**—The car does it all, assuming all safety-critical functions for the entire trip. The driver is not expected to control the vehicle at any time.

Figure 8.7 *The Navia Electronic Automated Transport, from Induct Technology.*

The ultimate self-driving car, then, is a Level 4 machine. We're at least ten years away from this ideal.

Introducing Google's Self-Driving Car

The most famous self-driving car today isn't commercially available, but it gets a lot of press. That's probably because it comes from Google, the big online search company.

Why Google wants to develop self-driving cars is another story, rooted in its drive to create street view photographs of seemingly the entire world for its Google Maps service. (Plus Google is decidedly anti-human in its use of technology; it'd rather make decisions by algorithm than leave it up to human beings.) Officially, Google says that its goal is "improving road safety and transforming mobility for millions of people." Sounds noble enough.

Google started out by retrofitting a variety of stock vehicles—Audi TT's, Lexus RS450s, and Toyota Priuses—with autonomous driving hardware and software. With that experience under its collective belt, Google then took the next logical step (because everything they do is exceedingly logical) and built its own prototype self-driving car. The company plans to build 200 or so of these robotic driving machines, all set to tool around California for the next year or so.

As you can see in Figure 8.8, Google's car (it has no formal name) is not a scion of automotive fashion. It's kind of cute, actually, especially with the perceived smiley face in the front. To me, it looks a lot like those Little Tykes plastic toy cars that my grandkids used to roll around in.

Figure 8.8   *Google's self-driving car, complete with hardware on top.*

It's not a toy, however. The Google car is a battery-powered electric vehicle, capable of a maximum speed of 25 mph. It has a stop/start button, but no steering wheel or pedals. There's room for two inside, but the body is plasticky.

Inside Google's car is a variety of sensors and technologies to detect the vehicle's surroundings. We're talking stereo cameras, 360-degree cameras, LIDAR, radar, and sonar devices. All these devices are necessary because they look at things differently; they all have different ranges and fields of view, thus serving a particular purpose in the grand plan.

For example, stereo cameras—two cameras mounted with a small separation between them—are mounted around the exterior of the car. They're used to create overlapping fields of views that can track an object's distance in real time. These cameras have a 50-degree field of view, but are only accurate up to about 30 meters.

More accurate distance is measured by the car's LIDAR system, mounted on the top of the car. This system is powered by a Velodyne 64-beam laser that can rotate 360 degrees and take up to 1.3 million readings per second. The LIDAR system is accurate up to 100 meters, which makes it ideal for generating a real-time map of the car's surroundings.

Radar systems are built into the car's front and back bumpers. These systems are used to warn of impending impacts—and tell the brakes to activate when necessary.

Interestingly, the radar systems are paired with sonar systems. That's because the two technologies are best for different distances. Radar works up to 200 meters away, while sonar is good for distances of 6 meters or less.

Up to 1GB of data is generated from these sensors every second. This data is used to build a map of the car's immediate surroundings. This enables the car to stay in the correct lane and avoid obstacles.

As the first company doing serious testing on self-driving cars in the wild, it's learning as it goes—which is a good thing. The more we know about how things work, the more effective (and safer) our smart cars can be. So far, Google's smart car has learned a lot about different driving conditions and hazards, including cyclist hand signals, railroad crossings, and what to do when another car is pulled over on the shoulder of the road. But, as any human driver can tell you, there's a lot more to learn.

Pros and Cons of Autonomous Autos

Why is everyone so excited about self-driving cars? Aside from being futuristic neat and all? Well, there are some definite benefits that accrue when you let your sedan or sport utility vehicle (SUV) do the driving—as well as some possible detriments.

The Good

Here are the good things made possible by self-driving cars:

- Fewer traffic collisions, due to accident-avoidance systems. Ideally, if all systems are working correctly, we should approach *zero* collisions.

That means less money spent on car repairs, medical expenses, and the like.

- Fewer collisions equal fewer injuries and fewer deaths. Today, around 35,000 people die every year in automobile accidents, and 90 percent of these crashes are due to human error. Smart self-driving cars that take the human factor out of the equation should dramatically reduce this accident rate.

- Given the lower accident rate, insurance rates should go down. (Should.)

- Reduced traffic congestion and increased roadway capacity, because autonomous vehicles can drive closer together without hitting each other.

- Reduced drive times, due to the higher speeds enabled by self-driving systems.

- Easier parking, because the car does it for you. In fact, the ultimate self-driving car will drop you off at the door and then go park itself.

- Longer drives without stops due to driver fatigue.

- A less stressful—and more productive—drive for occupants who no longer have to concentrate on driving. Instead, passengers can read, surf the Web, work, or just take a nap during the trip.

- Driver constraints removed. Since the car's doing the driving, it doesn't matter if the occupants are under age, over age, intoxicated, or blind.

- Less need for traffic police and vehicle insurance—assuming all the systems work to obey traffic laws and reduce the number of accidents.

- Reduction in car theft. A self-aware robotic car could just drive away if a bad guy tries to steal it.

- Lower costs for companies (formerly) employing human drivers. Domino's will save a ton on delivery expenses, which they might (or might not) pass on to customers.

Sounds great, doesn't it? Fewer accidents, fewer injuries and deaths, lower costs, easier drives. Sign me up!

 Note

You know who really likes the concept of self-driving cars? Uber, that's who. The crowdsourced transportation network/taxi replacement service intends to eventually replace all of its drivers with self-driving automobiles. Given that Uber's freelance drivers take home 75% of each fare, eliminating those drivers from the equation could be a financial boon to the company—and make it easier for people to find a ride, especially in big cities. On the downside, this would essentially destroy the current taxicab industry and put 10 million people out of work. Is this feasible? More than you think: A Columbia University study suggests that a fleet of just 9,000 autonomous cars could replace every taxi in New York City. The repercussions would be staggering.

The Bad

Of course, not all is milk and honey in the land of autonomous automobiles. There are more than a handful of potential downsides if you and your neighbors all have self-driving cars. These include the following:

- **Liability**—When your self-driving car gets in an accident, who's responsible—you or the car? (Or the car's manufacturer? Or systems programmer?) Nobody knows just yet.

- **Reliability**—Let's face it, if the computer in your car is as reliable as the computer on your desktop, we're all in trouble. Can you imagine having to reboot your vehicle in the middle of the drive to work?

- **Privacy**—Your smart car will collect a lot of information about you— where you drive, how fast you drive, and so forth. Who will have access to this data, and what will they do with it?

- **Security**—If a computer's in charge of your car, what do you do if that computer gets hacked? Vehicular cyberattackers could take control of your self-driving car, making it drive to strange places (where you could be robbed), or just stop working altogether. For that matter, less malicious hackers could retrieve your driving data and use it to send you targeted advertising. Just as you get with your PC.

- **Terrorism**—Cyberterrorists could program self-driving cars filled with bombs to initiate deadly attacks. It's not fiction; if you can imagine it, someone will try it.

- **Driver resistance**—Some people like to drive and don't want to cede control to some computer system. Won't self-driving cars take all the fun out of driving?

- **Loss of experience**—When your car does all the driving for you, what do you do when the situation requires you to take over the driving? If you're logging less time behind the wheel, your driving skills may deteriorate.

- **Loss of jobs**—Self-driving cars may be a good thing for employers, but could put a lot of minimum-wage drivers out of work. Say goodbye to the Domino's delivery guy and the taxi drivers in line at the airport. This could reshape the entire service economy.

Okay. That gives you something to think about.

The Ugly

Then there are the ethical questions. If a self-driving car senses that it's about to hit another vehicle, but swerving out of the way will cause it to crash into a pedestrian, what does it do? Given the choice of injuring a car's occupants in a collision or injuring the occupants of another car, what decision does it make? To what extent does your autonomous vehicle go to protect you—even if it means harming other drivers?

These are difficult choices made no less difficult when they involve algorithms and programming. And maybe you have different views on this subject than does your car's manufacturer.

Ethically, there are several different approaches that could be taken. Your car could be programmed to be

- Democratic, assuming that everyone in a given scenario has equal value.
- Pragmatic, so that certain people are judged more important than others. For example, the car might be programmed to give precedence to children in a school zone or pedestrians in city driving.
- Self-centered, so that you, the occupant of the host vehicle, always comes first.
- Materialistic, so that the least property damage (or legal liability) is inflicted.

While these ethical viewpoints could be programmed into the car by the manufacturer, it's just as likely that car owners will be allowed to determine which operational actions to take in the event of a pending accident. You, the car owner, could dial in just how much you want to protect yourself versus the occupants of other cars or pedestrians in the event of an accident. This shifts the ethical liability away from the manufacturer and onto you.

But how many people want to think about this sort of thing in advance? Would most people simply accept the manufacturer's default ethical settings? Would most people even read the obligatory legal disclaimer? (Probably not; do you ever read the Terms of Use that come with all the software you use today?)

This leads us not into ethical dilemmas, but also legal ones. If the car's manufacturer configures a self-driving car with an assortment of ethical "if then" programming, is the company then responsible for any injuries or deaths caused by the implementation of that programming? Oooh, you see how complicated this is going to be...

Navigating the Legal Landscape

The liability questions (and maybe the ethical ones, too) will eventually be decided by the courts and our lawmakers. Given the speed that our lawmakers work, however, the technology may get there before the laws can catch up.

The situation is this. If your self-driving gets into an accident (while in autonomous driving mode, that is), who is held legally responsible? Is it you, the car's owner—even if you weren't physically driving? Is it the car's manufacturer? Of the developer of the auto-driving software? Where does the proverbial buck stop?

Does the liability change if you, the driver, have some input as to how the car responds in a given situation? If the manufacturer programs a set of ethical responses (save the car's occupants first, or always avoid hitting pedestrians, or what not), does that put the legal liability in the lap of the car company?

And, before we even get to that point, just how legal is it to pilot a self-driving car on your local roads? Can you use a self-driving car today or not?

In the United States, at least, state vehicle laws typically do not envisage the adaption of self-driving cars. That creates a legal limbo. While existing laws don't necessarily prohibit autonomous vehicles, they do not explicitly allow them either. In fact, today's laws implicitly assume that a human being is behind the wheel and responsible for the car's actions.

With the looming introduction of self-driving cars on a mass scale, those laws will have to start changing. And they are. So far, only a handful of states—California, Florida, Michigan, Nevada, and the District of Columbia—have enacted laws that make the use of self-driving cars explicitly lawful, but many other states have similar legislation in the works. It's likely that self-driving cars will be explicitly permitted in most jurisdictions within the next several years.

These laws, however, do not address the legal liability issues, which are still in flux—and much debated. Motor vehicle departments and lawmakers across the United States are currently discussing this issue and writing draft legislation. Automakers, insurance companies, and other industry groups are offering their own helpful advice. And drivers like you and me will no doubt have our chance to sound off as well.

The point is, for self-driving cars to be successful—in fact, for manufacturers to introduce them for sale at all—the legal issues have to be decided in advance. No automaker with its staff of thousands of legal beagles is going to allow their cars to reach the market not knowing whether it will be legally responsible for whatever may happen out there on the road. Who's responsible for what and when will be determined and written into law, and the auto insurance industry will adapt coverage, policies, and rates accordingly. While it's all up in the air now, little will be left to chance by the time the big rollouts occur.

In other words, don't worry too much about this one. The big guys are working it out.

SMART CARS AND YOU

Today, smart cars—at least in the form of fully autonomous vehicles—are simply not available. They won't be for a number of years—probably two or more cars away in your auto-buying future.

So you don't have to worry about the ethics or legal liabilities of self-driving cars—not for a few years at least.

And when they do hit the market, they'll be available first from the luxury carmakers (Audi, BMW, Mercedes, and the like) not from your local Ford or Chevy (or Honda or Toyota) dealer. Like most automotive technologies, autonomous driving systems will migrate from the high end down to the low end, taking several years to make that journey. So the age of the $15,000 self-driving Kia is decades away.

That said, if you're in the market for a new car today, you do have some smart technology to consider. Higher-end models have the most "smart" options, as can be expected, but even the under-$20,000 mass market is seeing smart car technology creeping into the options list. My wife's new 2014 Honda Accord, for example, has a lane drifting alarm, backup and blind spot cameras, and all the Bluetooth-enabled entertainment you can dream of. As this technology becomes more affordable, you'll see more of it in more vehicles.

Smarter diagnostics are probably no more than a few years away, too. The totally self-diagnosing automobile, complete with the ability to phone into the mothership to schedule repairs, is still a futuristic concept. But the idea of more descriptive—and more proactive—in-dash warnings is something that's definitely on a shorter release schedule.

The point here is to be aware of the impending smart technologies (including all the self-driving stuff), know your options, and then take advantage of what's available when you're in the market for a new car. Some manufacturers are going to be ahead of others in this game, but they're all dipping their toes into the waters. Whatever smart technology you're waiting for will get here, eventually. Just be patient.

9

Smart Aircraft: Invasion of the Drones

What's a smart aircraft? One that can practically fly by itself. It doesn't have to be big, it doesn't have to carry passengers, it just has to fly by itself. You know what we're talking about here—in a word, drones.

Drone aircraft are all the rage today. We're fighting wars with drones. Your local police force is monitoring your neighborhood with drones. And the pizza joint down the street would like to deliver your next pepperoni pie with a drone.

Drones, drones, and more drones. There's a lot that these smart robotic aircraft can do once they take to the skies. Are you ready for the coming drone invasion? It's part and parcel of the Internet of Things (IoT), as you'll soon see.

What Drones Are—and What They Aren't

A drone aircraft is, nothing more and nothing less, an unmanned aerial vehicle, commonly referred to as a UAV. That means it's an aircraft, of any size or type, that flies by itself, without an onboard pilot or passengers. Think of it as a robot plane, controlled either autonomously or remotely.

Understanding Radio-Controlled Aircraft

Despite all the recent drone buzz, autonomous aircraft are nothing new. In fact, today's drones resemble nothing more and nothing less than the remote-controlled (r/c) airplanes and helicopters you used to fly as a kid. (Or maybe still fly; there's no age limit to the hobby.)

A typical r/c airplane, like the one in Figure 9.1, consists of an aircraft body, engine (gasoline, electric, or gas turbine), and some sort of wireless remote controller. Most airplane models are driven by propellers; some are powered by small jet engines. Models ranges from small and inexpensive to much larger (7-foot wingspans!) and much more expensive.

Figure 9.1   *The ready-to-fly UMX F4U Corsair r/c aircraft, from E-flite.*

Then there are r/c helicopters, like the one in Figure 9.2, which are a tad bit harder to fly but perhaps more versatile. Attach a digital still or video camera to the copter, control it with your smartphone, and you can take really cool pictures or videos from hundreds of feet in the air. That's probably why r/c helicopters are gaining in popularity among hobbyists and the general public.

Figure 9.2   *The m11 FHM Cobra Mini r/c helicopter, from Cobra RC Toys.*

The newest type of r/c aircraft, and perhaps the easiest to fly, is the quadrotor heli-copter, commonly called the *quadcopter*. As you can see in Figure 9.3, a quadcopter has four fixed pitched rotors; two rotate clockwise and two counterclockwise, pro-viding remarkable in-flight stability. Quadcopters are most like the drone aircraft currently employed in most civilian situations.

Figure 9.3   *The ELEV-8 V2 Quadcopter, from Parallax, Inc.*

You control your r/c aircraft with a handheld controller unit, like the one shown in Figure 9.4. Some "ready-to-fly" (RTF) models come with their own controllers; others include a radio receiver so you can use your own controller. All control signals are sent via radio frequency (RF) signals (in either the 72MHz or 2.4GHz bands), and let you control the craft's rudder, elevator, throttle, aileron, and more.

Figure 9.4   *The DX6 controller from Spektrum.*

You can purchase r/c planes and copters from Toys R Us or your local or online hobby shop. Some more toy-like models cost under $100, while true hobbyists will spend well in excess of that, particularly for models with more powerful engines, sophisticated controllers, and photo/video shooting capability.

What Makes a Drone a Drone?

So you spend a few hundred bucks for an r/c quadcopter or plane. Does that put you in the drone business?

Yes, it does. Since a drone is, by definition, any type of UAV, if you have an r/c aircraft, you have a drone. In fact, many r/c manufacturers are jumping on the bandwagon and re-labeling their hobbyist aircraft (especially those with built-in cameras) as drones.

As an example, consider the Parrot Bebop Drone, shown in Figure 9.5. (Yes, the word "drone" is right in the product's name.) This popular little quadcopter includes a built-in computerized global positioning system (GPS) navigation system that enables it to hover in flight and automatically return to its launch location. It also includes an onboard high-definition (HD) camera and creates its own Wi-Fi hotspot so you can control it with your smartphone or tablet, or its own optional controller. The Parrot's first-person view (FPV) capability means that you see what it sees in flight. The basic unit sells for $499.

Figure 9.5 *The Parrot Bebop Drone quadcopter in flight; note the camera mounted in the nose of the craft.*

 Note

R/c aircraft with FPV capability enable the controller to fly the craft beyond normal visual range. An FPV craft includes a video camera, typically mounted in the nose of the plane; the human controller views the live feed from the camera on a video screen. Some FPV craft have a range of 20 to 30 miles!

That said, today's more advanced drones are subtly and significantly different from hobbyist aircraft in a number of ways.

First of all, the most advanced drones today are moving toward autonomous or semi-autonomous operation. While external operator control is still common, automatic systems are increasingly being used to launch and land the craft, and to perform simple operations while in flight. Some drones employ GPS technology to enable autopilot modes.

Second, commercial and military drones are typically larger than hobbyist aircraft and often purpose-built. A drone used by the military to launch missiles at an enemy target is far removed from an r/c airplane designed for fun flying.

Third, commercial and military drones have much longer ranges than do hobbyist aircraft. They can not only fly further on a single charge or tank of gas, but they can also be controlled from a much greater distance. Whereas most r/c craft can fly only as far as the land-based pilot's visual range, commercial/military drones can be flown from thousands of miles away, using a combination of satellite and autonomous control technologies. This lets a home base remotely control drones across an entire city, state, country, or even continent.

Different Kinds of Drones

Speaking of drone control range, experts define drones in terms of how far away they can fly from their home base and how high they can fly. These classifications are detailed in Table 9.1.

Table 9.1   Drone Range Classification

Classification	Altitude	Control Range (max)
Handheld	600 meters (2,000 feet)	2 km
Close	1,500 meters (5,000 feet)	10 km
North Atlantic Treaty Organization (NATO) type	3,000 meters (10,000 feet)	500 km
Tactical	5,500 meters (18,000 feet)	160 km
Medium Altitude, Long Endurance (MALE)	9,000 meters (30,000 feet)	200 km
High Altitude, Long Endurance (HALE)	9,100 meters (30,000 feet)	Infinite
Hypersonic	15,200 meters (50,000 feet— suborbital altitude)	200 km

In addition to these range classifications, experts divide drones into six functional categories, as detailed in Table 9.2. (Note that some drones function in two or more categories simultaneously.)

Table 9.2   Drone Functional Classification

Category	Functions
Civil and Commercial	Provide functionality for non-military operations
Combat	Capable of attacking enemy installations
Logistics	Perform cargo hauling and other logistics functions
Reconnaissance	Provide intelligence information
Research and Development	Provide testing and other functions that further develop drone technologies
Target and Decoy	Identify targets to ground and aerial gunnery

How Drones Are Used Today

Aside from hobbyists flying what are in effect traditional r/c aircraft, there are lots of different entities employing drones today. More than 50 different countries—including the United States, China, and Iran—have their own drone programs, as do scores of state and local governments and police forces. An increasing number of private companies are also using drones for their own specific business purposes.

 Note

The drone business is booming, which is attracting lots of profit-hungry companies. There are more than a thousand companies in the drone industry, from the big military-industrial-intelligence complex players to smaller startups.

Military Drones

As you might suspect, the military has its own unique uses for drone aircraft. It's all about automating warfare, for good or bad.

The United States military is one of the, if not the, largest supporters of drone aircraft, to date deploying more than 11,000 drones. The military's drones carry out a variety of missions, from aerial reconnaissance to more controversial remote controlled combat. Reconnaissance drones are outfitted with HD cameras; combat drones are outfitted with missiles and bombs. Some experts believe that drones could eventually replace most manned military aircraft, with the corresponding savings of pilot lives.

The military likes drones for a number of reasons. First, they're a lot cheaper than traditional aircraft. Two, they can stay aloft longer than manned aircraft—several days at a time, in fact. And third, when one crashes, no personnel are hurt or killed or captured by enemy armies.

 Note

> While one hopes that that last reason is the driving factor behind military drone adoption, one also suspects that the dollar cost is of at least equal importance.

The most popular drone in the military's arsenal is the hulking MQ-1 Predator, from General Atomics, shown in Figure 9.6. The Predator, which has a 27-foot wingspan, is typically armed with AGM-114 Hellfire air-to-ground missiles. These are said to be especially effective at blowing up bad guys.

Figure 9.6   *The missile-equipped Predator drone in action.*

The Predators (and their larger Reaper cousins) are launched by ground crews near the conflict zone du jour, then operation is handed over to controllers 7,500 miles away at the Nellis and Creech Air Force bases in Nevada. There's actually a three-person control crew for each drone, each huddled in front of a bank of video screens. One person flies the drone, another monitors and operates the drone's cameras and sensors, and a third is in constant radio contact with the commanders and troops in the conflict zone. All attack decisions are made manually; there's nothing autonomous in the decision-making process.

Also popular is the considerably smaller (4.5-foot wingspan) Raven drone from AeroVironment, shown in Figure 9.7. The Raven is used for remote reconnaissance and is more autonomous than the Predator. After launch (which can be by hand), the drone gets its directions via GPS technology and reports back with a live video feed. These smaller drones can fly for days at a time without much if any human interaction.

Figure 9.7   *Launching the Raven reconnaissance drone.*

Intelligence Drones

The Air Force isn't the only U.S. agency using drones for military operations. The Central Intelligence Agency (CIA) is also a big supporter of drone technology. Where the military is somewhat overt in its use of drones, operating them where U.S. troops are deployed, the CIA is more covert, using r/c aircraft in situations that are below the radar, so to speak.

Not surprisingly, the CIA got into the drone business as a result of the 9/11 terrorist attacks. At the time, there was an increasing call for improved intelligence gathering, and drone aircraft fit the bill. The CIA's aircraft in Afghanistan, Pakistan, Yemen, Somalia, and other countries started out as a way to spy on suspected terrorist organizations. However, the spooks at the CIA soon graduated to more offensive operations, employing military-grade drones to bomb the bejeezus out of bad guys near and far.

To that end, the CIA uses drones in the Middle East and elsewhere to remove (re: assassinate) alleged terrorist leaders. And anyone in the nearby vicinity, because those Hellfire missiles aren't really that precise in their targeting. According to The Brookings Institution, for every drone attack that takes out a militant leader, ten civilians are also killed. This has led to the expected outrage from the international community, as well as from groups within the United States. Some view this targeted killing as a blatant violation of international law; others view it as the natural evolution of modern warfare, and one that actually saves the lives of those combatants who no longer have to be present for the killing.

 Note

> There's also the question of whether remote operators become trigger-happy because they're located in complete safety a continent or so away from the action. Or whether said operators suffer the same type of post-traumatic stress disorder (PTSD) often experienced by traditional combat troops.

Surveillance Drones

The CIA isn't the only organization using drones to spy on bad guys. The U.S. Border Patrol, for example, uses drones to patrol the nation's southern border, keeping a lookout for both illegal immigrants and drug traffickers. The agency favors both Predators and Ravens, and has a $39.4 million budget for aerial surveillance.

Then there's your local police force. Now, it's unlikely that the Danville, Illinois, police department is using missile-armed Predators to take out flagrant traffic offenders. But it is possible that they're using drones to spy on suspected criminals, monitor Special Weapons and Tactics (SWAT) operations, and maybe even keep an eye in the sky on traffic problems.

You can understand the appeal. A basic surveillance drone costs a lot less than a police helicopter, and can go places where copters can't. Drones can also stay in the air longer without refueling or recharging. What's not to like?

It's not surprising, then, to discover that close to a hundred local and state police agencies have applied to the Federal Aviation Administration (FAA) to deploy drone aircraft in their jurisdictions. Perhaps more surprising is the growing backlash against the domestic use of drone technology. Local and state lawmakers across the country have passed or proposed legislation severely limiting how and when law enforcement can use drone aircraft. For example, in Charlottesville, Virginia (the first city to place restrictions on drone usage), police are explicitly prohibited from using in criminal trials any information obtained by drones. Other jurisdictions—and even Congress, on a Federal level—have introduced legislation prohibiting drones from conducting targeted surveillance of individuals without a warrant.

So maybe your local cops are spying on you from the sky, and maybe they're not. It's an ongoing debate.

Civilian Drones

Drone use isn't limited to governments, local or national. Many, many organizations and companies are examining how they can employ drones in their day-to-day operations.

For example, energy companies are looking to use drones to inspect the oil and gas pipelines that crisscross the country. Electric companies are also considering drone surveillance of their power lines. (Drones are great for jobs that are too dull, dirty, or dangerous for manned aircraft or human inspection.)

Ranchers are using drones to monitor their livestock. The Bureau of Forestry is using drones to map wildfires, and to assist in search and rescue operations in wilderness areas.

Some real estate agencies are already using simple quadcopters to take aerial photographs of properties for sale. Retailers and food joints, from Amazon to Domino's, are evaluating drones for their delivery needs.

Hollywood is using drones to record footage for their films, as are big advertising agencies for footage for their commercials. Drones are being used to provide coverage for major league sporting events, including the 2014 Sochi Winter Olympics. And the musical group OK Go recently employed a drone to record the mind-boggling video for their song "I Won't Let You Down." (Check it out on YouTube; it's wild.)

There are also lots of scientific uses for drones. Drones are currently being used by the National Oceanic and Atmospheric Administration (NOAA) to monitor hurricanes and other meteorological disturbances. Some governments and research facilities use drones for scientific research, especially in severe climates like the Antarctic.

And there's more to come.

The Future of Drone Aircraft

Ten years ago, few people knew what a drone was. Today, drones are being used by governmental and civilian agencies for a variety of purposes. What will the world of drones look like a decade hence?

Smarter Drones

Many of today's drones really aren't "smart" aircraft. As noted, they're slightly evolved radio-controlled aircraft, controlled by operators either nearby or far away. There's not much autonomy in their operation.

Surveillance drones, on the other hand, are more autonomous. Once launched, today's surveillance drones, like the Raven, can pretty much operate on their own. They use GPS technology and computerized maps to find and hover near their targets, and automatically send back digital photos or a live video feed. There's not much the remote teams have to do, other than analyze the photos, videos, and other data sent back by the drone. If the drone needs to move to another location, the control team punches in the new coordinates, but then the drone does the rest.

As drone technology matures, more and more of the in-flight operations will become autonomous. Smarter drones will launch and land themselves, as well as find their targets for either surveillance or attack. More and better built-in sensors will enable drones to deal with real-time obstacles, in the form of weather, enemy fire, even other aircraft in the area. Today, the remote control team needs to pilot drones out of any pending trouble; tomorrow, drones will be able to cope with obstacles themselves.

It's also possible that combat drones will be able to make autonomous attack decisions. Today, it's up to someone on the control team to pull the figurative trigger, but smarter targeting software will enable future drones to make those decisions without human interaction. The drone will use facial recognition software to identify key targets, identify other people and property in the area, calculate when the target is acquired and when there's an acceptable risk of collateral damage, and then, when everything checks out, fire the missiles. No need to subject human beings to these difficult decisions (and resulting PTSD); just let the robots do the killing for us.

If this sounds a little too much like the *Terminator* scenario for comfort, consider that at a much smaller scale, these kinds of automated attack decisions are already being made. Israel's Harpy Unmanned Combat Air Vehicle is programmed to recognize and automatically dive bomb any radar signal that isn't in its built-in database of friendly sources. That enables it to locate and destroy enemy anti-aircraft radar installations, for example—all without any humans involved in the decision.

Note

Technology seldom stays limited to the country that produced it. To that end, Israel has already sold its Harpy drone to China, India, South Korea, and other countries.

Autonomous weapons of all sorts are especially troubling. Do we really want robot troops fighting our wars for us? On the one hand, it'll cut down on our side's casualties. On the other hand... Well, there's that whole *Terminator* thing to worry about. And even if Skynet never comes into being, the morality of automating killing in combat is suspect.

Whether you find robotic combat—either in the air or on the ground—troubling or exciting, it's something that needs further discussion. Of which, find more in Chapter 10, "Smart Warfare: Rise of the Machines."

Delivery Drones

Smarter drones don't have just military applications. The more intelligent a drone is, in terms of directions and flight, the more useful it can be in a variety of civilian applications.

Probably the most talked about commercial application for smart drones is in product delivery. Right now, delivery is a big cost center for businesses that sell things online or over the phone; it's also a major expense for food delivery businesses. Whether we're talking a decently paid United Parcel Service (UPS) or FedEx driver, or a minimum wage pizza delivery guy, businesses would love to lower their delivery costs—and automating delivery is one way to do that.

It's not surprising, then, that businesses as diverse as Amazon and Domino's are all evaluating the potential use of drone aircraft to deliver their products. One relatively low-cost drone could replace not only current delivery personnel but also their trucks and cars. It's also possible that drones might be faster and more accurate than the high school kid with a stack of pizzas in the back seat.

For drones to be effective delivery vehicles, they have to be smarter than they are today. Your local Chinese restaurant isn't going to employ a team of drone pilots jiggling joysticks in the joint's back room; businesses want to input the delivery address, load up the drone with their product, and let it fly. There's no sense using drones if you just replace one human employee (the delivery guy) with another (the presumably higher-paid drone pilot).

This points to the need for smarter navigation systems. The drone has to know where a given address is, the best route there, and any potential obstacles in the path. Fortunately, these are all available technologies.

Consider the example of the pizza delivery drone. You call in your order, or place it online, and it's entered into the system at the pizza joint. When the pizza slides out of the oven, it's boxed up and carried over to the drone launch area. The pizza is affixed to the bottom of the next drone in line, which is then fed the delivery address from the store's computer system. The appropriate button is pushed, and the drone lifts off into the night.

The delivery coordinates are in the drone's computer memory, along with aerial maps of the area. The drone can fly pretty much straight-line to your address, although tall buildings and power lines are part of the mapping system, so the drone knows to avoid these types of obstacles. The drone also is equipped with collision avoidance systems, so if there are any other drones (or birds or low-flying aircraft) in the area, it will adjust its flight path as necessary.

When the drone arrives at your address, it lands or otherwise drops the pizza on your front doorstep, and texts you that your delivery is ready. You've already paid in advance via credit card, of course, so all you have to do is open the front door, retrieve your pizza, and wave goodbye to the drone as it flies back to home base.

If this whole scenario sounds a bit farfetched, know that back in 2013, Domino's in the United Kingdom (UK) tested their own DomiCopter, as they called it. As you can see in Figure 9.8, this was a six-rotor UAV adapted to carry one of Domino's insulated pizza pouches. Now, the DomiCopter wasn't much more than a PR stunt designed to cash in on the ongoing dronemania, but it still was a nice proof of concept for drone delivery.

Figure 9.8 *Domino's six-rotor test DomiCopter in flight.*

Also testing the drone delivery waters is Amazon, the big online retailer. Amazon is developing what it calls Prime Air, a drone-based delivery service that promises 30-minute delivery times in select locations. As shown in Figure 9.9, the Prime Air drone is an eight-rotor UAV (technically, an octocopter), with enough lift to carry small packages. These battery-powered drones are said to fly at 50 mph with a 5-pound payload, with the necessary sensors and systems to avoid mid-air collisions. Amazon would like to launch Prime Air within the next year or so, although that probably won't happen due to FAA regulations. (Amazon says that 86 percent of the products in its inventory weigh five pounds or less, making them prime candidates for Prime Air drone delivery.)

Figure 9.9   *Amazon's Prime Air drone, carrying a test package.*

 Note

Amazon's Prime Air is designed for local deliveries. For longer-haul deliveries, consider the concept of the *dronenet*. This is essentially a network of drones that carry packages the same way the Internet carries data—in packets, over a series of multiple hops, with exact routes devised on the fly, so to speak. One drone delivers to another drone and thus to another, with a final local drone making the delivery to the specified home or office building.

Then there's Google, with its Project Wing. Developed at the company's top-secret Google X labs, Project Wing is a prototype unmanned delivery vehicle with unique vertical take-off and landing capabilities. This drone sits on its tail and has four rotors—two on the underside and two on the outside, toward the edges of the wings. It's not a helicopter and it's not a fixed-wing aircraft; it's a unique hybrid craft that combines elements of both approaches.

The aerodynamic design isn't the only thing unique about Project Wing. Google has been paying special attention to how the craft delivers its payload. Landing a drone in an urban or residential neighborhood is problematic, with lots of obstacles (both architectural and human). Google has posited other ways to drop a package from above and settled on winching the package down to the doorstop. The drone hovers in a fixed position while a line is reeled out, lowering the package to the ground, as shown in Figure 9.10. After the package lands, the line detaches and winds itself back into the craft. This eliminates unwanted human interaction with the craft. (Google found that with traditional landing delivery, too many test subjects reached out for the package while it was still attached to the drone—and risked injury from the drone's rotors.)

Figure 9.10 *Google's unique Project Wing drone, lowering a package for delivery.*

 Note

As appealing as drone delivery is for restaurants and retailers, there are other less commercial and more humanitarian applications for the technology. There are plenty of areas in Africa and other less-developed areas where roads are few and land travel is difficult. In these areas, drones could be used for more efficient and speedier delivery of essential supplies, including food and medicine. Similar applications exist for disaster relief.

Regulating Drone Aircraft

The big stumbling block for widespread commercial use of drone aircraft isn't technology, which is advancing just fine, thank you. It's regulation—and deciding who gets to use the friendly skies and how.

Regulation of the airspace in the United States is the responsibility of the FAA. The FAA typically deals with private and commercial aircraft and their flights through the nation's airways. But drones would fly in those same airways, and there's the rub.

Current FAA regulations severely restrict the commercial use of drone aircraft, even for research and development purposes. Amazon and Domino's just can't launch their drones into the wild blue yonder without the FAA's approval, and the agency is loathe to give carte blanche approval to what could amount to tens of thousands of commercial drones zigzagging across major cities. The opportunity for collision—with buildings, with other drones, with bigger aircraft, or with people (during landing or takeoff)—is just too great.

Current FAA regulations permit hobbyist drone or r/c aircraft use only when flown below 400 feet and within the operator's line of sight; anything beyond that is prohibited. Hobbyist aircraft also cannot be flown near airports or other zones with heavy air traffic.

These regulations don't always keep hobbyists from flying their r/c aircraft where they shouldn't. The FAA notes up to 25 cases each month of drones reported to be flying above the 400-foot limit, with some flying as high as 2,000 feet in the air. That's certainly high enough to interfere with commercial flights and represents a growing danger.

How is the FAA dealing with this threat? By increasing their education programs, that's how. And, boy, that's sure to work. All they have to do is let more people know that they shouldn't be flying their r/c craft so high or near airports, and the problem will go away. We'll see how that works.

 Note

> To its credit, the FAA is actively working with the Model Aeronautics Association and other groups and clubs in the model aircraft community to increase awareness levels.

That still leaves the issue of commercial drones. The business community wants to use them, but the professional aeronautics community and the FAA aren't quite so hot on the idea. Pilots in particular are concerned that these small r/c aircraft are difficult to see, and that their operators aren't properly trained in interacting with larger aircraft. (Pilots operate on the principle of "see and avoid," where they take proactive action to avoid other craft in the sky; drones can't really "see" other craft and may be too small themselves to be seen.)

Given the many benefits of drone technology, however, the FAA can't sit on its hands forever. The agency will be forced to allow the commercial use of drone aircraft; how it regulates said flights, however, is what we don't know.

The FAA could throw caution to the wind and deregulate all flights by aircraft under a certain size and weight. That's unlikely, and perhaps even unwise, but it's certainly an approach that Amazon and Domino's and their ilk would support.

Another approach is to create specified air corridors or layers of airspace for drone use. This way we'd know where the drones were flying, and other aircraft could stay the hell out of their way.

It's also possible that the FAA might require transponders or other signaling devices on commercial drone aircraft. This would enable drones to be tracked, which has multiple benefits. (Such as knowing when a drone flies somewhere it shouldn't.)

For the time being, however, drone delivery awaits FAA approval. That said, it's likely that there will be some action from the FAA in this area within the next year or so. Amazon and other big lobbyists will certainly be pressuring them to do so.

Fly the Scary Skies: The Problems with Drones

As you've no doubt sensed, there is some skepticism and trepidation about the continued deployment of drone technology. There are obviously benefits from using UAVs, but also some risks involved.

Collision and Liability Concerns

It's tempting to look up into the big blue sky and see it as a wide open playground. It's not. There are lots things sharing the airspace, from commercial aircraft to private aircraft to experimental aircraft to birds to balloons to fireworks to other drones. As more and more things launch themselves into the air, the risk of two or more of them colliding increases.

Now, if you're piloting a drone, the last thing you want is for it to run into something else in flight. This would, in most cases, quickly end the flight, and not much good comes when an object suddenly drops from a height to the ground. (It's not the drop that gets you—it's that stop at the end.)

And it's not just the drone that's at risk. Imagine a wayward drone crashing into a schoolyard playground filled with children. Or a drone flying into the jet engine of a commercial airliner. The results could be disastrous.

Obviously, drone manufacturers and operators want to minimize the risk of in-air collisions. This is being accomplished via the use of intelligent collision avoidance systems, employing radar and similar technologies to identify in-air obstacles and then steering around them.

While in-air collisions can be minimized, they can't be eliminated entirely. Nor can the risk of mechanical failure or weather-related situations, both of which can also cause a rapid fall from the previously friendly skies.

To that end, we come to the issue of who's responsible when a drone crashes, either into another flying object, the ground, or an unlucky bystander. If an Amazon delivery drone smashes through the roof of your house, who pays for the damages? Amazon, the drone manufacturer, the person piloting the drone, or somebody else?

These are issues that need to be addressed by all concerned parties, including the FAA and (inevitably) the insurance industry. Rest assured, there will be drone-related accidents, and they will cause a big hoodoo in the press, and there will be calls for action to do something about the growing menace from the skies. Getting in front of the issue is necessary.

Security Concerns

Here's an issue that's common to most devices in the Internet of Things. Since an autonomous drone aircraft will have its own onboard computer, and since computers can be hacked, what's to stop the criminal or mischievous element from hacking into and gaining control of that Amazon or Domino's drone?

The possibilities are endless. A terrorist group could hijack a drone and fly it into a civilian target. A teenaged hacker could hijack a drone and claim it as his own new plaything. A tech-savvy criminal could hijack a delivery drone and steal the merchandise being delivered. You get the idea.

Drone security needs to be addressed at the same level as IT-based computer security. No company or organization wants its drones hacked or hijacked, so precautions must be taken in advance to mitigate this risk. In-flight operations must be secure, or everyone is at risk.

Privacy Concerns

It's not surprising that the manufacturers of military-grade drones are looking to expand into the civilian market, and are targeting police forces and governments in cities and states across the United States. It's also not surprising that local authorities are chomping at the bit to get their hands on their own surveillance drones, which they feel can make them more effective at a lower cost than current methods.

Not everyone is enthusiastic about increasing the number of eyes in the skies, however. Many privacy advocates fear that this will dramatically expand the surveillance state, making it all too easy for your local, state, and national government to spy on anyone and everyone they like. (Or dislike, rather.)

Think about all the information that could be collected by a surveillance drone. We're talking high-resolution photographs, clear enough to read license plates and identify human targets. HD video that tracks a subject's actions over an extended time frame. Infrared and RF sensors that can peer through foliage and even buildings to detect people inside.

In short, drones can spy on just about anyone anywhere, from a far enough distance that that person will never know it. Sure, the government will only use these birds to track the bad guys—or so they'd have you believe. But what if the government mistakes *you* for a bad guy, or even thinks that you *are* a bad guy? Do you really want your every movement tracked by a drone hovering overhead?

The privacy threat is significant enough that big privacy groups are weighing in. The American Civil Liberties Union (ACLU), for example, is concerned that as drone aircraft become cheaper and more widely available, law enforcement agencies may be tempted to carry out persistent surveillance of U.S. citizens.

This is definitely something to be worried about. Just how much surveillance do we want to accept in our society?

Other Smart Aircraft Technologies

Drones aren't the only aircraft employing smart technologies. IoT-related tech is being employed in aircraft of all shapes and sizes, including both military and commercial craft.

Smart Structures

One such emerging technology is that of *smart structures*. These are structures—such as an aircraft's frame or covering—that can adapt to environmental conditions. In other words, smart structures sense their environment and then self-diagnose and adapt to the current condition.

To do this, smart structures employ a combination of technologies and disciplines, including materials science, sensors, actuators, nanotechnology, cybernetics, artificial intelligence, and something called biomimetics.

 Note

Biomimetics (sometimes called biomimicry) is the imitation or mimicry of the models, systems, and elements of nature.

Smart structure technology will enable aircraft manufacturers to reduce the aircraft's total weight, manufacturing cost, and operational costs by integrating a variety of system tasks into the structure itself. We're talking aircraft (or parts of aircraft) that can morph into different shapes over the course of a flight.

For example, instead of using fixed geometry wings, as is standard today, smart structure technology will enable the use of wings that subtly change shape during different parts of a flight. The wing might shape itself in a way to decrease noise during takeoff and landing, and then reshape itself to decrease drag (and thus increase both airspeed and fuel efficiency) during high-altitude flight.

Smart Skin

Similar to smart structure is the concept of *smart skin*. This is a covering applied to the outside of an aircraft that enables planes to sense their environment, via the use of thousands of tiny sensors, each no larger than a grain of sand, embedded into the plane's skin.

Smart skin technology will sense changes in temperature, wind speed, and the like, and then feed that information back to the plane's main computer. The plane can use this information to adjust flight speed, altitude, and similar parameters to minimize both flight time and fuel usage.

Smart Maintenance

Smart skin technology can also help aircraft monitor their own health and trigger necessary maintenance. Smart skin will report potential problems to engineers and maintenance crews, who can then do what they need to do to fix any potential problems.

Other smart technology will also help in problem diagnosis and ongoing aircraft maintenance. Expect future aircraft to have integrated structural health monitoring (SHM) systems to provide more timely notice of issues and reduce the cost of both inspections and repairs.

Smart Cabins

Smart technology will also be applied inside the cabins of commercial aircraft. Expect to see shape-changing seats that adapt to different body types (and budgets, with more space for higher-paying passengers), white noise and "sound shower" technologies that wrap passengers in private sound cocoons, and a bevy of new entertainment and communications options.

One company evaluating such smart cabins is aircraft manufacturer Airbus. The Airbus Concept Cabin, shown in Figure 9.11, breaks away from traditional Economy, Business, and First Class sections to zones within the plane for specific needs—relaxing, working, conducting business meetings, working, playing games, and so forth. The cabin adapts to each passenger's needs and provides a unique experience for each.

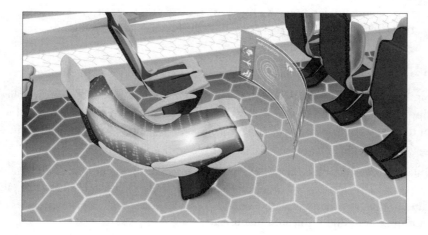

Figure 9.11   *The bio-morphing smart seat in Airbus' Concept Cabin of the future.*

For example, Figure 9.11 shows the Concept Cabin's bio-morphing seats that mold comfortably to each individual's shape and size. These seats will be made of lighter, thinner materials that provide more leg space and fit more passengers into the cabin. Sounds grand, even if the more dominant trend is cramming more passengers into an increasingly smaller and uncomfortable cabin space. We can dream, can't we?

SMART AIRCRAFT AND YOU

So how does all this smart aircraft development impact you?

While there's not much you need to do to prepare for the coming onslaught of drones and other smart aircraft, it's likely you'll benefit from the ongoing technological developments. Aside from the increasing risk of having a drone land on your head, there's not much downside for the average consumer.

Consider drone delivery. In spite of the FAA dragging its collective feet on this issue, it's going to happen and it's going to happen relatively soon. The issues, safety and otherwise, are known and can be worked out to everyone's benefit. Which means that sometime in the near future, you'll order a sandwich from Jimmy John's and have a drone drop it on your doorstep in 10 minutes or less. Basic orders from Amazon and other retailers, at least in large cities, will be delivered via drone. Anything light enough will fly your way via your friendly neighborhood drone.

This is all to the good. You'll get what you want faster and maybe even for a lower cost; drones cost less than human delivery people and their associated motorized vehicles. At some point, ten or fifteen years down the pike, we'll be wondering how we ever got along without drone delivery.

Of course, all these drones delivering pizzas and panties and who knows what else will contribute to a more crowded airspace. Especially in urban areas, you'll have to get used to the buzz of electric drones and the sight of dozens of these little buggers crisscrossing the skies overhead. You'll probably also get used to the sight of the occasional drone lying in a plastic heap on the sidewalk, having misjudged its delivery point and crashed into the side of a nearby building. It'll happen.

You'll also get used to not having a delivery guy knocking at your front door. No more hunky brown-shirted UPS guy, no more pimply high school pizza delivery guy. No more delivery guys, period. For you, that's fewer people to tip; for the businesses involved, that's fewer people to pay; for the former delivery employees, that's less work available. That's the way automation works.

The other thing you'll have to get used to is the way drones affect our armed forces and our country's combat operations. As we'll discuss in the next chapter, drones and other robotic vehicles will reduce the need for human troops and thus reduce combat deaths and casualties. This may also make it easier for us to wage war, which probably is not a good thing—even if fewer of our young men and women get killed in the process. Smart technology is going to change warfare in the same way that mechanized vehicles changed things during World War I. Technology marches on—even and especially across the battlefields of Earth.

10

Smart Warfare: Rise of the Machines

Missile-equipped drones. Smart bullets. Laser guns. Robot soldiers.

While we're not yet living in the science fiction world of self-aware killer robots, the face of combat is certainly changing. We have the Internet of Things (IoT) to thank—or blame—for that.

The Past, Present, Future of Tech-Based Warfare

Science fiction has imagined a variety of dystopic futures. That seems to be part and parcel of the genre. Malevolent robots, giant mechanical killing machines, particle-beam weapons—that sort of thing. The type of battlefield you see in *Terminator* and *Star Wars*—combat where human beings are little more than cannon fodder for the cyborgs and robotic weapons controlled from afar by digital commanders.

Technology has always influenced warfare and vice versa; much technological development has arisen during times of war. From catapults and cannons to jet fighters and atomic bombs, tech and combat go hand in hand.

Three Generations of Warfare

Military experts define three distinct generations of warfare, with a fourth generation looming on the horizon. Each generation is defined by the technologies and strategies employed.

The first generation of warfare was that of line-and-column combat. This is the type of combat common up to and including the Napoleonic wars, where masses of soldiers on both sides of the conflict lined up (in those rows and columns), marched toward each other, and used the technology of the day to try to kill each other. In early days, the technology was sticks and stones, then swords and maces, then muskets and bayonets. It was a brutal, bloody, personal business, and the available weapons technology really didn't help much.

The second generation saw the introduction of long distant battles, due in part to the development of longer range, more accurate guns and artillery. With these more advanced weapons, soldiers didn't have to stand right in front of the enemy to kill them; killing could now be done at a distance. This is the type of combat experienced during World War I. Again, the soldiers lined up in rows and columns, but instead of marching toward each other, they dug themselves into trenches and let their new technology weapons do the killing for them. On the front lines, automatic rifles and machine guns made killing much more efficient than old-tech weapons. Behind the front lines, killing could be accomplished from afar with cannons and other artillery. Throw a little mustard gas into the mix, along with the first aerial combat vehicles, and you get slaughter on a scale previously unimaginable.

In the third generation of warfare, combatants employed speed and surprise to bypass the enemy's lines and collapse their forces from the rear. These tactics were made possible due to the introduction of tanks and aircraft into the battle sphere. This is the type of combat employed during World War II and all subsequent wars, up to and including the 2003 American invasion of Iraq. More distance combat, more bombing and shelling, less face-to-face or hand-to-hand fighting—culminating in what we then thought was the ultimate bomb, like the ones dropped on Hiroshima and Nagasaki.

Notice the trend of relying more and more on machinery and technology, distancing fighting from individual soldiers. Combat becomes more impersonal, more a flick of the switch or push of a button than anything directly physical.

The Fourth Generation

This leads us into the coming fourth generation of warfare, where killing is done completely by remote control, with a lessened need for hand-to-hand combat. Attacks are more targeted, whether in terms of property or people. Drones and other smart weaponry are programmed with specific targets, and that's what they hit. There's less mass bombing and more pinpoint missions. Targets are determined using data from high-tech surveillance devices; commanders are made smarter, leading to more informed decisions (and, presumably, fewer mass civilian casualties).

When soldiers are employed, they're better protected and better armed. High-tech materials make for uniforms or protective shells that withstand enemy fire. Smart weapons enable precise targeting from a safe distance. State-of-the-art communications systems enable instantaneous communications between command and field, and video feeds enable commanders to see what the soldiers see in combat.

Of course, all of this military tech costs money, so fourth generation warfare is only affordable by the richest combatants. This puts poorer countries at a distinct disadvantage, but that's been ever the case. Victory belongs to those who can afford it.

Unless, that is, the ever-decreasing cost of technology actually levels the playing field. A small remote-controlled drone aircraft capable of carrying a decent-sized payload costs a lot less money than a state-of-the-art stealth bomber. Even otherwise-low-tech insurgents have access to all the information that is on the Internet, along with the corresponding real-time, long-distance communications. Robot soldiers may still be too expensive for non-superpowers, but targeted killing machines, made available via smart technology, may quickly get into the hands of even the smallest groups of enemy combatants.

Note

Warfare today is often *asymmetric*, where the military power of one side differs significantly from that of the other. This is typified by a traditional army fighting a smaller group of less-well-equipped insurgents. Give those insurgents some smart weapons, however, and the odds become even.

Smart Aircraft

Let's start our examination of smart military technology with smart aircraft, in the form of unmanned aerial vehicles or drones. Drone aircraft, more than any other current combat technology, is changing the nature of warfare today.

We talked in depth about drones in Chapter 9, "Smart Aircraft: Invasion of the Drones." There's no need to repeat all that information here, so turn back if you want to learn more about Predators, Ravens, and other types of drone aircraft used by the military.

With that basic information in mind, know that military around the world see drone technology as a significant part of future conflicts, both large and small. More than 70 countries currently have drones, and drones are being used today for both reconnaissance and targeted combat missions.

The United States first used armed drones in 1995, during the Yugoslavia Civil War. In the post-9/11 era, Predator MQ-1 and more advanced MQ-9 Reaper drones have been used to kill (some would say assassinate) Al Qaeda officials and other suspected terrorists in Pakistan, Afghanistan, Somalia, and other areas of conflict.

And it's not just the United States. Britain and Israel have used armed drones to target insurgents, Turkey is hoping to buy drones to fight Kurdish separatists, and even Iraq is investing in drones (made in the United States) to protect their oil platforms.

The biggest, baddest armed drone today is the MQ-9 Reaper from General Atomics, a San Diego-based defense contractor. The Reaper, shown in Figure 10.1, is a "hunter-killer" drone, its dual missions united to locate and then eradicate its targets. It has a 66-foot wingspan, which makes it more like a traditional aircraft in size than a radio-controlled toy plane. It's powered by a 950-shaft-horsepower turboprop engine, has a maximum speed of about 300 mph, has a range of 3,200 nautical miles, and can fly at an altitude of 50,000 feet. Reapers carry either Hellfire missiles or more intelligent laser-guided bombs, making them especially deadly, even as they're piloted from afar.

Figure 10.1   *The MQ-9 Reaper hunter-killer drone.*

Deadly as they are, drones are considered by many to be more humane than more traditional missiles and bombers. They conform more closely to the Geneva Convention, in that their more targeted systems supposedly minimize collateral damage. This last point is certainly debatable; a powerful missile delivered from a Reaper or Predator is certainly big enough to wipe out more than a single person walking down the street.

Some see drones as the future of military conflict. By precision targeting opposition leaders, buildings, and equipment, large-scale military operations can be minimized or eliminated completely. Drones also eliminate friendly casualties; there are no pilots or crew to be shot down over enemy territory. And if they truly do the job of stopping conflicts before they turn massive, ground-troop casualties will also be significantly reduced.

Others see drones as doing nothing more than shifting combat responsibility from the battlefield to the headquarters (HQ)—and, potentially, from men to machines. If human troops aren't getting their hands (as) dirty on the battlefield, if troops and commanders alike are not subject to the undeniable horrors of war, will war become more common? It's not surprising that some of the people who object most to sending troops into unnecessary combat are those who've been there themselves. It's more often military leadership that drags its heels on potential combat, while the politicians back home get all hawkish and patriotic. If our military leaders no longer have to worry about sending young men and women into dangerous situations, it's possible that trigger-happy politicians will lead us into more conflicts. Maybe it's just too easy to kill someone by pressing a button; maybe it needs to be more difficult than that.

 Note

Drone strikes are not as precise as the military would like you to believe. Reprieve, a British nonprofit organization, notes that, on average, every drone strike in the Middle East kills 28 bystanders for every intended target. In the so-far fruitless quest to kill Al Qaeda head Ayman al-Zawahiri, 105 innocent individuals have been killed, 76 of them children. Traditional attacks or "targeted" drone strikes, collateral damage—in the form of innocent lives lost—still exists.

Smart Bombs

Smart military aircraft become more dangerous when they carry more intelligent and more powerful payloads.

Today's largest drone, the MQ-9 Reaper, is big enough to carry all manner of traditional ordinance, including Hellfire missiles and 500-pound, laser-guided bombs. That's well and good, but what if the same (or smaller) craft were loaded with nanoexplosives that delivered twice the explosive power in the same package?

Nanoexplosives use nanoparticles that have more surface area in contact between the different chemicals used in the explosive. When a reaction is initiated, this extra surface area causes a faster reaction rate, which makes for a more powerful explosion. The upshot is that you can make a given-sized bomb twice as powerful, or get the same explosive force with a bomb only half as large—the latter being an attractive option when dealing with smaller and lighter drones.

As to smarter bombs, your tax dollars are funding research into this as well. A smart bomb, also called a guided bomb, is one that uses precision guidance systems to hone in on the intended target. These weapons are theoretically capable of being guided—either autonomously or via remote operator—directly to the target, thus ensuring more effective operation with less collateral damage.

These smart munitions, such as Boeing's GPU-31 Joint Direct Attack Munition (JDAM) and Raytheon's GBU-39 Small-Diameter Bomb (SDB), are similar to drone aircraft, in that they're unmanned and employ intelligent systems to guide the weapons to their targets. The JDAM employs global positioning system (GPS) technology for more precise targeting, while the SDB, shown in Figure 10.2, uses a combination of radar, infrared, and laser signals for guidance.

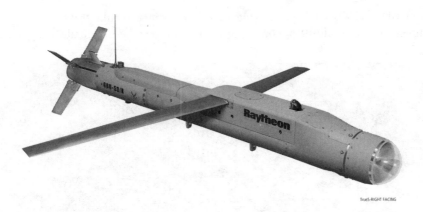

TrueS-RIGHT FACING

Figure 10.2   *Raytheon's SDB smart bomb.*

What you get with both these systems is a bomb that is guaranteed to hit its target, and not something a mile or so away. Traditional bombs are notoriously dumb; you drop them out of a plane and assume they hit something valuable when they reach the ground. Smart bombs, like the JDAM and SDB, can be directed at very precise targets and hit them. We've all seen the grainy video footage of a smart bomb in the first Iraq War aiming for and then barreling down the smokestack of a targeted building. The technology enables a higher success rate and fewer near misses.

 Note

> Because both of these weapons use radio frequency (RF) technology, they can be confused and thrown off course with radio-jamming technology. Which is why the Air Force is also working on what is called *home-on-jam* technology. When a jamming signal is detected, the home-on-jam technology takes over and flies the weapon directly to the source of the signal. Thus, the jamming signal becomes the opponent's demise.

Smart Weapons

Smart technology isn't limited to larger munitions. There's a whole generation of smart personal combat weapons either here now or on its way. We're talking smart guns and smart bullets, folks.

For example, the XM25 Counter Defilade Target Engagement (CDTE) System, shown in Figure 10.3, is a smart sniper's rifle (designed by Alliant Tech Systems and Heckler and Koch) that uses smart bullets programmed to explode when they've traveled a set distance. This enables the sniper to either hit the target directly or flush out the opponent, by exploding above or beside the bad guy.

Figure 10.3 *The XM25 CDTE smart rifle.*

Here's how it works.

The soldier shines a laser rangefinder at the enemy combatant, which then calculates the distance to the target. The bullet can be programmed to hit the target directly or to explode above or beside the target. Each bullet has a timed fuse, as well as a small magnetic transducer that interacts with the Earth's magnetic field. This transducer generates a small alternating current every time it spins, and the miniature computer inside the bullet counts the number of rotations it makes. When the bullet has flown the designated distance, the computer issues the instruction to detonate, releasing a burst of shrapnel.

Future weapons will enable soldiers to hit targets up to 1,000 yards away without any specialized training. This is due to developing technology from TrackingPoint, an Austin-based company specializing in smart rifle technology. A TrackingPoint system pairs a Linux-powered scope with a guided trigger; not only can distant targets be dialed in with a minimal amount of effort or skill, a live feed of what the shooter sees can be fed back to commanding officers. To give you an example of how effective this technology can be, a typical soldier only has 20 percent accuracy with targets 1,000 yards distant; with TrackingPoint technology, the accuracy jumps to 70 percent. With this kind of smart shooting, you don't need trained snipers to make the kill.

Then there's the new type of weapon called a *railgun*. This is an electromagnetic projectile launcher, capable of accelerating heavy projectiles at hypersonic speed. It's built around a pair of parallel conducting rails, over which a sliding armature is accelerated via the electromagnetic effects of a current that flows down one rail, into the armature, and then back along the other rail. This flings the armature, carrying an explosive payload, down the rail and into the air.

A railgun has several advantages over a traditional battlefield missile. First, you don't need any propellant; the missile flies off the rail, propelled solely by the electromagnetic pulse. Second, the missile doesn't need to be packed with high explosives; it's traveling so fast (up to Mach 7—seven times the speed of sound) that its speed alone makes it highly lethal. Third, railgun systems employ guided projectiles, so they're extremely accurate.

Railguns are well into the developmental phase. For example, General Atomics' Blitzer railgun, shown in Figure 10.4, is expected to be ready for production by 2016. It can be used either on land or on the sea, to knock out enemy battleships.

Figure 10.4   *The Blitzer railgun in action.*

There's more. The U.S. military is working on real honest-to-goodness laser guns that look more like *Star Trek* than anything in use today. Particle-beam weapons will focus a high-energy beam of subatomic particles to disrupt the target's molecular structure. And electromagnetic pulse weapons will be capable of knocking out all of an enemy's electrical systems, from computers to car engines, without destroying life or property.

Fascinating and scary stuff, brought to you courtesy of your local military-industrial complex. That said, these weapons promise to make combat more deadly for the enemy, while at the same time safer for troops on our side.

Robot Soldiers

If you want to get into real science fiction territory, consider the robotization of combat. We're talking everything from mechanical combat suits to full-blown battlefield robots. Robotic battlefield technology promises to make combat more survivable for the grunts on the ground, as well as reduce the number of soldiers needed to fight.

Sound far-fetched? Then talk to General Robert Cone, Commander of the Army Training and Doctrine Command (TRADOC). He's looking at shrinking the size of the combat brigade from about 4,000 soldiers to 3,000 and filling the gap with robots.

That's right. Robots.

 Note

Experts note that replacing human soldiers with robots will save lives (at least on our side). The motivation isn't all humanitarian, however. Building robot soldiers, at least in the long run, will be cheaper than training, feeding, and caring for human soldiers. It's an economic decision as well as a strategic one.

Today's Army Robots

Let's start our look at robotic warfare by looking at the robots employed by the U.S. Army today. There are a few—although most are rather primitive in their construction and functionality.

First, there are *bomb-squad robots*. These little robotic devices were used in both Iraq and Afghanistan to dispose of improvised explosive devices (IEDs), those deadly roadside bombs. Better to have a robot try to diffuse an IED—and possibly get blown up doing so—than risk the life of a human member of the bomb squad.

The Army has also tested several remote-controlled, machine gun-firing robots. One of the more promising units is the Protector, from HDT Global Dynamics. As you can see in Figure 10.5, the Protector is a small (three-foot wide) tank-like unit with an M240 machine gun mounted on top. All the operating mechanisms are protected in the tank part of the unit, which can carry 1,350 pounds of gear. The diesel-powered, 32-horsepower engine enables it to climb 45-degree slopes with ease.

Figure 10.5 *HDT's Protector machine gun-firing robot.*

The Protector is operated via wireless remote control, up to 3,280 feet away. The operator uses a controller that looks a lot like a videogame controller, with two buttons and a thumbstick. It also includes a "cruise control" button that enables the robot to prowl on its own power, maintaining the current speed and direction. It can also be equipped with various attachments for other duties, including a backhoe and mine roller/rake.

Then there's the Legged Squad Support System, or LS3, a robotic pack mule from Boston Dynamics (now a Google company). As you can see in Figure 10.6, the LS3 looks a little like the All Terrain Armored Transport (AT-AT) walkers in the *Star Wars* movies, but on a considerably smaller scale.

The LS3 has four legs and walks kind of like a dog. (Check YouTube for a video; it's kind of creepily fascinating.) It can carry 400 pounds of supplies, follow squad members through rugged terrain, and interact with troops in a semi-autonomous fashion, much like a trained animal interacts with its handler.

 Note

The LS3 is nicknamed AlphaDog, and follows on its predecessor, the aptly named BigDog.

Figure 10.6   *The LS3 at work.*

There are presently several autonomy settings in the LS3's programming which affect how it follows along with its squad. For example, in Leader-Follower Tight mode, the LS3 attempts to follow its leader as closely as possible. In Leader-Follower Corridor mode, the LS3 sticks to the leader but has the freedom to make localized path decisions. In Go-to-Waypoint mode, the LS3 ignores the leader and uses its sensors to avoid obstacles on the way to a designated coordinate.

The whole point of the LS3 system is to relieve troops of some of the burden of carrying supplies. This will help troops move faster and have more endurance in the field.

Robotic Armor and Super Soldiers

Human beings are regrettably fragile. Not only do we Homo sapiens tire quickly and require 8 hours or so of rest each day, we're also quite vulnerable to damage from bullets, grenades, bombs, and similar things. That makes for high numbers of casualties in combat, which is a bad deal not only for those who are injured or killed, but also for their commanders who must continue replenishing the ranks with fresh troops.

What if there were a way to make humans a little less vulnerable? To enhance not only their survivability but also their abilities? What if we could use technology to create an army of super soldiers, capable of marching longer without rest, suffering fewer injuries, and having more and stronger offensive powers?

Not surprisingly, the U.S. government is working on just such a thing. And the technology is almost there.

The simplest(!) approach is to encase soldiers in robotic armor. We're talking some sort of powered mechanical exoskeleton (for enhanced strength and protection) combined with wearable computers, communications gear, and smart weapons.

The exoskeleton is powered by a system of motors and hydraulics that enhance the wearer's strength, powering at least part of the limb movement. This will enable the wearer to carry heavier loads both in and out of combat. The exoskeleton will also help the wearer survive in dangerous environments—and absorb the shock of bullets and explosions. It's kind of like the Iron Man suit, but without Tony Stark inside.

The military has commissioned several prototypes of powered exoskeletons, including the Human Universal Load Carrier (HULC—get it?) from Lockheed Martin and the XOS from Raytheon. The XOS, shown in Figure 10.7, weighs about 200 pounds, is constructed from high-strength aluminum and steel, and uses a variety of controllers, sensors, and actuators to perform necessary tasks. The powered limbs enable the wearer to lift more than 200 pounds of weight without feeling any strain.

Figure 10.7   *Raytheon's XOS military-grade-powered exoskeleton.*

The exoskeleton is for protection and enhanced strength. It needs to be accompanied by some sort of control mechanism, probably mounted in the soldier's helmet. Consider such a helmet with accompanying visor that contains a heads-up virtual display, night-vision capability, and wireless headset and microphone. The soldier turns to look at a target, locks onto that target with the heads up display, says "Fire!," and sees the target go down in front of him.

The next step is to augment soldiers' bodies with robotic technology. This would turn the average soldier into a bit of a cyborg—a "cybernetic organism," if you will. This approach, while definitely science fiction-y, promises the best of both worlds; you get the durability, strength, and precision of robotic systems combined with the superior cognitive abilities of the human brain.

This is serious stuff here. We're talking a combination of neuroscience, nanomedicine, robotics, computer technology, and a whole lot more. Tapping into soldiers' brains so they can control their robotic armor and weapons systems through the power of thought. Melding mechanical systems with biological ones. Creating a super soldier whose mind and body are one with his armor and weapons.

Such a cyborg soldier will be capable of fighting in all manner of extreme environments, from the cold of Siberia to the heat of the Iraqi desert. He will have the endurance to remain in the field for extended periods of time. Embedded nanosensors will constantly monitor his medical status, nanoneedles will release drugs when needed, and nanorobots will quickly heal his wounds. Optical implants will enable him to see in the darkest night, to view the enemy via infrared light, and extend his vision thousands of feet. He will be a super man.

Our future cyborg soldier will still need his robotic armor, of course. Embedded into this armor will be advanced intelligent weapons systems. A rapid-fire machine gun melded onto one arm, a precision long-range sniper's rifle or even a laser rifle on the other. Shoulder-mounted missile launchers. A stock of smart bullets and smart missiles ready for automatic reloading. The necessary communications systems embedded into his helmet. Everything he will need in the field attached to his armor. A one man fighting machine, controlled by that one man's human intelligence.

Yes, it sounds far-fetched. But it's all in development today.

Take, for example, the Cognitive Technology Threat Warning System (CT2WS), a helmet-mounted, threat-detection system, shown in Figure 10.8. This system consists of three parts. The first part is the electroencephalogram (EEG) headset that uses wireless sensors to monitor electrical activity in the brain, looking for a specific brainwave that results when the subconscious detects a visual threat. The second part is an electro-optical video camera with a 120-degree field of view that scans the surroundings. And the third part is the computer system, to which both the EEG unit and camera are connected; it uses proprietary algorithms to assess the threat level and identify potential targets. Initial tests show that this system identifies 91 percent of enemy targets, as opposed to just 41 percent identified by nonaugmented humans.

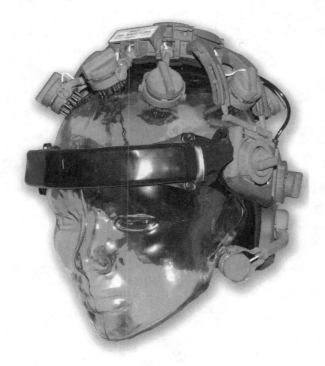

Figure 10.8   *The CT2WS threat detection headset.*

And there's more to come. Any volunteers?

Autonomous Fighting Robots

We're a long way from developing cyborg soldiers to do our fighting for us. The progress is slow, even with the backing of the Defense Advanced Research Projects Agency (DARPA—the same folks who developed the Internet). There are all sorts of hurdles to overcome, from computing power to the necessary portable power sources to all the neuro-biological issues. And even when we develop the perfect super soldier, there will still be a human being inside—and human beings are still human, after all.

The ultimate fighting machine, then, will probably be just that—a machine. Our pals at DARPA have been funding robot research for years, with the goal of creating a robot army that can do our fighting for us, via either remote control or autonomous operation.

Consider DARPA's Atlas robot, shown in Figure 10.9, created by Boston Dynamics (the same folks behind the AlphaDog). Atlas is 6 foot 2 inches tall, humanoid in design, and is a biped—that is, he walks on two feet, like you and me. This leaves his arms free to lift and carry things, as well as perform other tasks with his articulated robotic hands.

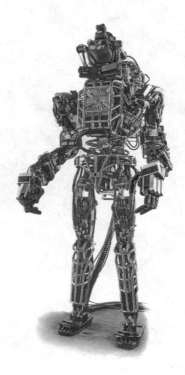

Figure 10.9   *The Atlas robot.*

Atlas was purportedly developed for disaster response and currently sports no built-in weapons systems. But it's not a stretch to see this big guy adapted for combat use at some time in the future. It's what the folks at DARPA tend to do.

Current technology has a long way to go, however, for robots to become effective battlefield fighters. In particular, the robots have to get a lot smarter so that they can perform autonomously in the field; one simply can't imagine a battalion of 4,000 marching robots being remotely controlled by 4,000 human operators. The robots have to "think" on their own and adapt to the ever-changing conditions of combat.

But we don't want them to get *too* smart. The idea of robots developing sentience is the stuff of science fiction (think the Skynet system in the *Terminator* movies), but a very real possibility. We really don't want our killing machines gaining self-awareness and staging some sort of cybernetic revolt. This means putting some sort of limits on artificial intelligence technology, or embedding strong and effective constraints on the machines' autonomy circuits. Or something like that.

 Note

> The point where a computer's artificial intelligence exceeds human intellectual capacity (and control) is called the *singularity event*. Resulting events beyond this singularity are impossible to predict, but probably won't be good for humankind—do we really want our machines to be smarter than we are?

Smart Strategy

Perhaps more important than smart weapons and cybernetic soldiers is the decision-making that is crucial to all combat operations. The more information military commanders have, the better the decisions they'll make.

Here is where the Internet of Things will make a real difference in how future wars are conducted. The IoT is all about smart things—and collecting massive amounts of data.

Information gathering and analysis is the key to future warfare. Imagine everything combat troops experienced beamed back in real-time to HQ. Imagine satellites transmitting high-resolution images in real time. Imagine information-gathering drones and ground robots, feeding their data back to HQ. Imagine insect-sized devices flying into the enemy zone, undetected, each sending small amounts of data that can be combined and collated to give commanders a near-perfect view of what the opposition is up to.

Now imagine the intelligent systems capable of receiving, analyzing, and acting on all this data. Systems that can identify and summarize the important data for the generals to act on. Systems that can even act autonomously on the data, making decisions without the involvement of the military brass. Battles conducted based on data and logic rather than human emotion. Preemptive strikes that halt combat before it starts.

It all starts to look a little like some computer war game. Maybe humans won't be involved at all—our computers will battle their computers in some virtual space, and we'll get a text message informing us who won. Why not? It's got to be better than the way we conduct our wars today.

SMART COMBAT AND YOU

Hopefully, the evolution of smart combat will have very little impact on you as a civilian, except perhaps to make you a little more safe. For those in the military, however, there will likely be significant impact.

The United States military—and, in fact, the entire defense industry—has been aware of and interested in the Internet of Things long before it had that name. In what the military used to call (and sometimes still calls) the "connected battlefield," all the airborne, seaborne, and land vehicles, all the weapons systems, all the units on the battlefield are networked together, to share and act on tactical information.

The military's concept of the connected battlefield (or IoT combat) is constantly expanding as more and newer technologies come to the fore. Autonomous vehicles present new possibilities for both lightening troop burden and advancing combat effectiveness. Embedded and ruggedized computers add intelligence and memory to military equipment large and small. Sensor devices help to gather and process large amounts of combat-related data. In-the-field network solutions (routers, switches, and such) provide the necessary connectivity infrastructure. More robust computers and systems process the big data generated by all these devices and help commanders make more informed strategic decisions.

The goal is to connect every battlefield asset, large or small—every soldier, every vehicle, every weapon, every device—and then use those connections for more effective decision-making and operations. Way back in the 1990s, the Department of Defense (DoD) identified four key elements of the connected battlefield of the future:

- Networked forces with improved information sharing
- Information sharing and collaboration that enhance quality and situational awareness
- Shared situational awareness that enables self-synchronization (the process of coordinating units on the battlefield)
- The combination of the first three elements to increase mission effectiveness

Today, in 2015, much of this vision is now a reality. Forces are increasingly networked and sharing key information; that collaboration is enhancing the troops' situational awareness; this is enabling battlefield units to better coordinate their actions; and everything is working in concert to increase the effectiveness of ongoing military missions. Units and troops on the field of combat are empowered to make more decisions in real time, without waiting for decisions and approvals from above. This results in faster tactical decisions and swifter, more effective operations.

In short, and as always, technological advances create distinct advantages in the field of battle. That said, this coming round of IoT-related advances also holds potential problems, especially if the killing decisions are taken out of human control.

The issues are many. Will soldiers be willing to work alongside autonomous robots? Will soldiers agree to modify their own bodies for the sake of the cause? Will robots one day completely replace human soldiers and human-operated systems? Will commanders trust the machines in their control? Will all the devices on the battlefield return reliable data? And can we trust our computers to act appropriately on that data?

Finally, can we be sure that the smart systems we develop won't turn against us? It's entirely possible that artificial intelligence systems will develop in ways not yet anticipated. What if an intelligent combat robot decides that the best way to achieve its mission is to fire on its own troops? What if some massive computer intelligence decides that human beings are superfluous—or that warfare itself is illogical? If and when the singularity is achieved, who knows what will happen?

So, while it's unlikely that we civilians will be directly affected by the integration of the IoT into the military, we may be indirectly affected by the decisions made afterwards. As informed citizens, we should stay abreast of these new technologies and how the military plans to use them; even if all they do is make warfare a little easier, that may not be something we want as a society.

11

Smart Medicine: We Have the Technology...

With self-monitoring medical devices, barcode prescription tracking, interconnected devices in hospitals, and cloud-based medical recordkeeping, the world of the Six Million Dollar Man isn't far off. Why visit a doctor when your own medical implants can diagnose things automatically? And it's all connected, so if something goes wrong, your implants can schedule a doctor's appointment—or call 911—if you need help.

It's a given that the Internet of Things (IoT) is going to change healthcare as we know it. By connecting together all the various medical devices currently in use (or soon to be introduced), healthcare gets a lot smarter real fast. Diagnoses come quicker and are more accurate, mistakes are fewer, and patients get better care and more effective preventive care. It's a medical dream come true.

Welcome to the Internet of Medical Things

Of all the areas that the IoT promises to revolutionize, the one that stands out from the rest is healthcare. The infusion of smart connected technology into the world of medicine is changing what happens when we get sick or injured—for the better.

Connecting Devices

Most of this change occurs when existing medical devices are connected together. We're talking all the different devices in your hospital room, in the intensive care unit (ICU), in the emergency room (ER), and in the operating room (OR). Plus those medical devices you might have on your person at home—wearable or implanted blood pressure (BP) monitors, heartbeat monitors, even pacemakers. All these devices can be connected wirelessly to the IoT, and then to each other or to a central database or monitoring system. Instead of reporting independently, these devices send data in real time to an automated or doctor-monitored service that can make immediate sense out of all the disparate readings—and even initiate actions (such as prescribing new meds) based on that analysis.

In other words, you don't have to wait until your next visit to the doctor's office to tell him that you're feeling under the weather, have him draw blood and run the requisite tests, and then get back to you with a new prescription. Thanks to the IoT, all this information is available much sooner, in a combined form that makes sense of it all, and in a way that enables near-immediate analysis and response. This is how healthcare is revolutionized.

 Note

Some consider medical applications to be a distinct subset of the Internet of Things, sometimes referred to as the Internet of Things for Medical Devices, or IoT-MD.

Centralizing Records

Additional IoT-fueled change comes from the centralization of medical records. Right now, your records are entered by hand and kept (probably on paper) at your doctor's office. If you have two or more doctors for different conditions, they each keep their own records. Even if you visit multiple doctors at the same facility, chances are they don't share their records—or even talk to each other about you. And if you end up at urgent care or the ER, that's another set of records that your regular physician doesn't know about.

The IoT is changing all that. Physical records are being digitized, and new records are entered into centralized computer databases. Data is stored, not in a single office (or multiple offices) but in the cloud, so that all your various physicians, clinics, and hospitals can view and contribute to them as necessary. One doctor will be able to see what another has prescribed, eliminating cross-diagnoses and dangerous drug interactions, and resulting in more coordinated, effective, and efficient treatment.

This should also reduce medical costs, at least over the long run. If and when the right diagnosis is arrived at more rapidly, there will be less waste in the system. In addition, the IoT promises to make preventive medicine more common, and healthier people should require less longer-term medical attention.

In short, we're looking at nothing less than a real revolution in healthcare. The results will be a lot more useful than that smart TV or talking refrigerators you've been eyeing.

Realizing Benefits

There's a lot involved in cobbling together the IoT in the healthcare environment, but also some real and significant benefits from doing so. At the end of the day, why bother with the IoT-MD? Here's why:

- **Decreased cost of care**—Big benefit—the IoT promises to lower healthcare costs. We need that.

- **Improved patient outcomes**—Put another way, patients get better quicker, and with fewer mistakes along the way. If you're sick, you're more likely to get well.

- **Real-time disease management**—Instead of waiting for your quarterly appointment to check your blood sugar or thyroid levels, they get monitored every single day—and the results fed back to your doctor so changes in medication can be determined. In fact, the real-time, constant monitoring inherent in the IoT may reduce the number of doctor visits you need.

- **Improved quality of life**—With more effective treatments and more preventive treatments, people with both transitory and chronic illnesses will live more normal lives. In this instance, normal is good.

- **Improved user experience**—This means for both the patient and the healthcare providers. Yes, you want to improve the experience for the person receiving the healthcare, but there's also benefit in making things easier and more user-friendly for the doctors, nurses, and assistants doing the day-to-day heavy lifting. If an IoT-based system can cut their workload or make it easier to make decisions or perform certain operations, all to the good.

Smart Medical Devices and Monitoring

One of the first and most important steps to smarter healthcare is the use of remote monitoring devices. These are already becoming popular, due to an increased focus on preventive care and readmission prevention. (In fact, we've discussed some of these devices previously, such as fitness bands and heart rate monitors, in Chapter 6, "Smart Clothing: Wearable Tech.")

These smart medical devices enable patients to receive the same level of monitoring at home that was previously only available in hospitals and other medical facilities. We're talking remote monitoring wherever you may go, enabled by wireless communication over the Internet, that enables your physician or healthcare facility to track your health stats, physical activity, and drug consumption in real time— and alert the necessary personnel if a medical emergency occurs.

Examining Smart Medical Devices

What types of connected medical devices are we talking about? Here's a short list:

- Blood glucose monitors that track your blood sugar levels
- BP monitors that provide real-time BP readings
- Breathing monitors that track your pulmonary ventilation
- Electrocardiogram (ECG) monitors that track cardiac activity
- Electroencephalogram (EEG) headsets that keep tabs on your brain activity
- Hearing aids that tailor their performance to room conditions—and connect, via Bluetooth, to your smartphone so you can use them as headsets, as well
- Heart rate monitors that track the number of beats per minute, such as the Mio Link Heart Rate Wristband, shown in Figure 11.1
- Muscle fatigue monitors that use muscle contraction sensors to help in fitness training
- Pacemakers that send back information on the wearer's heart rate and condition
- Pregnancy monitors to track the unborn baby's heart rate and other key stats

Figure 11.1   *Monitor your continuous heart rate with the Mio Link Heart Rate Wristband and accompanying smartphone app.*

- Pulse oximeters that monitor pulse and blood oxygen (O2) saturation via a fingertip sensor
- Sleep monitors to track and evaluate sleep patterns and sleep apnea
- Stress monitors (such as the Spire, shown in Figure 11.2) that track motion and respiratory patterns to determine when a person is becoming stressed or unfocused

Data from these individual devices can be integrated into larger systems (more on that in a moment) or tracked via PC/smartphone apps. For example, Manipal Health Services in India provides pregnant mothers with a wearable fetal monitoring system that ties into a corresponding smartphone app. This system provides mothers with real-time data about fetal heart rate patterns, uterine activities, and labor progress. This data is also transmitted to the physician's smartphone or tablet; he or she can then make more informed decisions about what to do next. (The system is even active and useful during labor and delivery.)

And there's more to come. How about micro cameras, in the form of edible pills, that enable doctors to remotely observe your internal conditions without the risk of surgery? Or an embedded motion detector that doctors can use to monitor the symptoms of Parkinson's disease? Or a tattoo-like skin patch, like the one in Figure 11.3, that contains flexible, stretchable electronic sensors to monitor heart, brain, and muscle activity—or stimulate muscles, remotely—in a noninvasive manner?

Figure 11.2 *The Spire stress monitor clips onto your belt or shirt and lets you know when you're getting a little tense.*

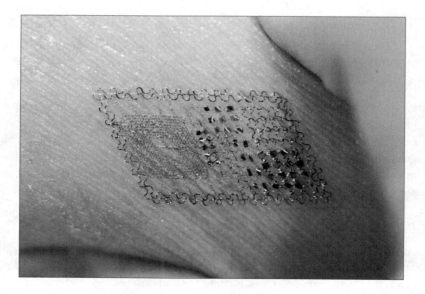

Figure 11.3 *Monitor all sorts of vitals with this micro-electronic skin patch, developed at the University of Illinois at Urbana-Champaign.*

With all these devices connected to a central monitoring system, doctors will be able to quickly and easily access patient information—both historical and current. This data can be combined with information gathered by other connected devices, such as smart scales, fitness bands, smartwatches, and the like, to provide a more complete picture of the patient's health.

All this sounds futuristic, and it is. That's because we're marching inexorably into the future, where the IoT reigns. Behold the future—it is upon us!

Monitoring the Monitors

Once medical data is collected, we need software that looks for trends in the patient's stats. Obvious anomalies are easily noted (and can trigger alerts), but the real value comes in detecting long-term changes in the patient's health. It's not just the continuous monitoring; it's how that data can inform health-related decisions and actions.

Ideally, this type of ongoing monitoring and analysis can help keep you from needing emergency or extraordinary care. If you have BP issues, for example, these remote systems will see it and suggest the proper medication before it becomes a life-threatening problem.

The challenge is dealing with the vast amounts of data collected. It's not nearly as simple as an individual checking his or her progress on a smartphone app. We're talking data from multiple devices at multiple times—for each and every person in the medical registry. If and when everything gets connected and the data gets sent to the appropriate systems and services, making sense of these huge amounts of data is a mind-boggling enterprise.

It's also a huge opportunity to those companies that can figure it out and offer solutions to hospitals and other healthcare providers. One company that's working on it is Freescale Semiconductor. Freescale has developed a reference platform they call the Home Health Hub that enables application developers to create "telehealth" applications that collect and share patient data. As you can see in Figure 11.4, the Home Health Hub works with commercially available healthcare devices (BP monitors, thermometers, pulse oximeters, scales, blood glucose monitors, and the like), then distributes the collected data over the cloud to remote smart devices, such as PCs, smartphones, and tablets. This enables physicians, caregivers, and others a way to monitor the patient's current health status, as well as provide alerts and medication reminders.

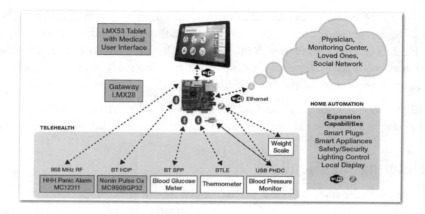

Figure 11.4 *Freescale's Home Health Hub reference platform ties together multiple medical devices and beams the results to a mobile app. (Diagram courtesy Freescale Semiconductor, Inc.)*

To make sure that most, if not all, medical devices work together to share their data in this fashion, the Continua Health Alliance, a coalition of healthcare and technology companies, has established guidelines for what they dub interoperable personal health solutions. These specs enable conforming devices to work together over Bluetooth, Wi-Fi, and ZigBee networks to share and distribute data. This specification is the backbone for the IoT in healthcare applications.

 Note

In addition, the U.S. Food and Drug Administration (FDA) has recognized 25 standards that support medical device interoperability.

Smart Monitoring for Seniors

The medical IoT holds particular promise for seniors, who often need the type of monitoring available today only in hospitals or assisted living facilities. Newer wireless monitors and reporting systems will enable seniors to stay in their homes and still receive the quality of care they need.

Some of this monitoring can be accomplished by the types of wearable devices we've been discussing. In addition, specialized sensors can be installed in the homes or apartments of older patients to monitor their health and physical activities.

This may sound futuristic, but much of this functionality is available today. For example, Healthsense offers the eNeighbor monitoring system for seniors. It uses sensors worn by the patient and installed throughout the home to detect any falls, note if an elderly patient has wandered off, and alert caregivers if the patient has missed taking any medication. The system also includes an emergency call pendant, which seniors can use to manually call for help if necessary.

Similarly, the BeClose system places smart sensors, like those shown in Figure 11.5, around the patient's home to keep track of normal routines. When something is amiss—a prolonged absence, for example, or a missed meal—a caregiver or designated family member is notified by either text message, email, or phone call. If installed in a senior living facility, staff can monitor each patient's activities via an interactive dashboard. (And still receive necessary alerts when things aren't as expected, of course.)

Figure 11.5   *The BeClose system uses a variety of smart sensors to monitor seniors' activities.*

Then there's Independa, which offers Remote Care technology that can be used in either assisted living facilities or in seniors' homes. All monitored data is displayed on the Caregiver Dashboard, shown in Figure 11.6, which can be viewed by caregivers over the Web on any personal computer, smartphone, tablet, or smart TV. The Dashboard also displays "smart reminders" for medications, events, and check-ins, as well as a Wellness Summary, suggested health measures, reports, and alerts.

There are many other companies getting into connected senior care, including a lot of custom proprietary solutions. The goal is the same for each—to improve the quality of life for seniors, while providing effective monitoring to keep them safe and healthy.

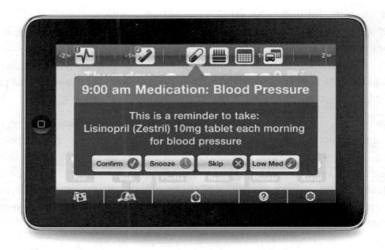

Figure 11.6 *Independa's Caregiver Dashboard displayed on an iPad.*

Smart Meds

Here's a frightening fact: Every year, more than a million people in the United States suffer from medication errors—taking the wrong meds or the wrong quantity of meds. These errors should be almost completely preventable—and may well be, thanks to the Internet of Things.

There are numerous approaches in development that will help ensure that people like you and me (and our elderly parents and grandparents) take only those drugs that have been prescribed and in the correct dosages. It's a great application of developing technology.

The first and easiest solution is to use barcodes on all medicine containers. This type of barcode-based medication administration system is in use in many hospitals today, and they've cut the error rate substantially. This type of system is a bit unwieldy, however, and is typically tied to a workstation on wheels—not practical for in-home use.

 Note

Barcoded prescription bottles also streamline the process of reordering and refilling medications. Patients scan their old bottles with a smartphone app that automatically places the refill order with their pharmacy. This is happening today; Walgreens reports that 40 percent of their customers reorder their meds using the company's smartphone app.

An alternative approach is to make the whole process wireless. Medication containers have embedded radio-frequency identification (RFID) tags which are read (using near field communication [NFC] technology) by a tablet or smartphone. It's more portable than the old barcode system and offers the potential for additional uses, such as tracking whether or not a given drug was beneficial to the patient. It's all a matter of recording and storing the proper data, and then forwarding that data to the appropriate analytical application.

When used at home, an RFID-based medication tracking system can help to confirm that patients are actually taking their medications and remind them to do so. Data collected from the mobile app will be beamed back to the physician's office, which can then generate summary reports or contact the patient with a gentle reminder or warning to do what they're supposed to do.

For example, Vitality's GlowCap, shown in Figure 11.7, is an electronic cap you can fit on just about any prescription drug bottle. It uses light and sound reminders to signal when it's time to take that particular medication. When the pill bottle is opened, a chip inside the GlowCap wirelessly relays that information to your physician or caregiver. (GlowCap uses AT&T's Mobile Broadband Network for data transmission, hence that rather large round wireless transmitter plugged into the wall in the background.) When all the pills are gone, push a button at the base of the lid to order a refill.

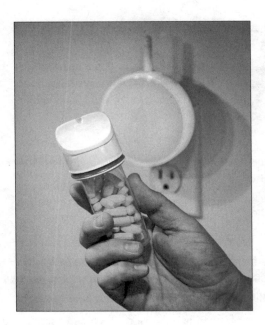

Figure 11.7   *GlowCaps remind patients when it's time to take their pills—and notifies physicians when they have.*

The next step moves beyond batch identification built into the meds' container to identifying each unique pill in a bottle. If all pills are manufactured with built-in RFID transmitters, then they can be tracked individually via smartphone or tablet apps.

There's a lot of data that could be gathered by smart pills. For example, the Digital Health Feedback System from Proteus features an ingestible sensor placed within a smart pill. The sensor is powered by contact with your stomach and transmits information not only about when the pill was swallowed, but also provides metrics regarding heart rate, body position, and activity. In this instance, the smart pill beams its data to a micro-electronic patch on the user's stomach. The patch then transmits the data via Bluetooth to the patient's smartphone, and from there to the doctor or other healthcare provider. The whole system is shown in Figure 11.8.

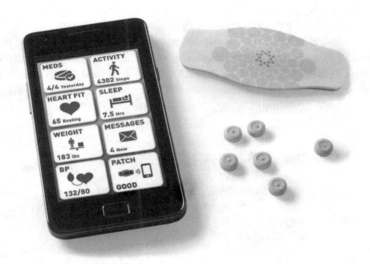

Figure 11.8 *The smart pills, transmitter patch, and smartphone app that comprise Proteus' Digital Health Feedback System.*

Smart Hospitals

Hospital error is the sixth leading cause of preventable death in the United States. More than 50,000 people are killed every year in the United States alone due to something going wrong during a hospital stay.

Human beings being human beings, we're all going to screw up from time to time. But there are ways to reduce human error—and simultaneously improve the quality of hospital care.

Cue the Internet of Things and related smart technology. Connected devices and systems, along with intelligent data distribution and analysis, promise a revolution in the way hospitals work.

Everything's Monitored, and Nothing's Monitored

This won't be the first time that technology has radically changed patient care. Thirty or so years ago, hospitals realized that continuously monitoring patients dramatically improved outcomes. This led to a revolution in patient monitoring (and the medical device industry), with the resulting placement of ECG monitors, pulse oximeters, and multi-parameter monitors in every hospital room. Today, patients are hooked up to a half-dozen or so machines that track how they breathe, how their heart beats, how high their BP is, and more—and, if they're on any intravenous drugs, a different infusion pump for each medication. If something goes haywire, the beeping starts and you get a nurse or doctor rushing in to take care of things.

With all these monitors already in place, what can possibly go wrong? Lots, actually. That's because in most instances, these devices are not connected to each other or to any central monitoring facility. They all operate independently.

And they're location-dependent. If a patient moves from one room to another, or from the ICU or ER to a private room, he has to be disconnected from all the devices in the first room and reconnected to the devices in the next one. The devices in the first room don't talk to the devices in the other room, so there's no electronic history recorded or transmitted. All the data is unique to a particular device.

In addition, there's no way to combine the data from these different devices to generate a holistic report of patient status. You end up getting a lot of nuisance alarms, fatigued staff dealing with all these different units, and a lack of understanding about what's really happening with a patient's health. And nobody is looking at this data in real time—only when something goes wrong.

My stepdaughter is a neonatal care nurse at a local hospital. She vouches for the inefficiency and ineffectiveness of today's multiple-monitor hospital room. Yes, all these monitors are for the good, alerting staff when something goes off the norm. But there's no way to look at all the monitor data in one place, either in the room or at the nurse's station. And, believe it or not, long-term data is not stored or transmitted from any single device; if a nurse or doctor wants a record of a given statistic, she has to manually make a chart. That's manually, as in by hand, using pencil and paper.

Wouldn't it be better if all these devices not only talked to each other, but also to some sort of centralized dashboard? Where the settings from a machine in the

ICU, OR, or ER could be automatically transmitted to the devices in the patient's hospital room? Where a report or chart could be generated with a push of a button—even combining data collected from multiple devices?

Yes, it would. And that's where the Internet of Things comes in.

Smarter Devices

It's really a matter of connectivity. All the devices used by a given patient need to be connected to one another and to some sort of central control/monitoring station. That would make all these devices smarter and ultimately more effective—and reduce the amount of time that staff spend dealing with them.

What you want is for all settings to follow a patient, no matter where he or she is in the hospital. If a given infusion pump knows that the patient needs a certain amount of a given drug, then the infusion pump in the next room should be able to pick up those instructions without manual programming by the staff. (Each staff interaction introduces the possibility of human error, of course.) The historical data from the heart monitor in the first room should be combined with the more recent data collected from the similar monitor in the next room. And all this data should be displayed in real time, in summary and detailed fashion, on monitors in the patient's room and at the nurse's station.

Even better, you want all the connected devices to work together to produce smarter results. Consider the oximeter, which monitors the patient's O2 levels. Today, low oximeter readings cause a lot of alarms, even though low O2 levels are typically only a concern when accompanied by a low respiratory rate. A better approach, then, would be a smart alarm that checks both O2 and carbon dioxide (CO_2) levels. But that's two different monitoring devices today, and they don't talk to each other. Enable cross-device communication and some intelligent if-this-then-this algorithms, and you get a much more effective alarm—and fewer false alarms.

This doesn't sound difficult, but it is a major challenge. In any given hospital, it's a matter of dealing with, at any given time, thousands of patients and hundreds of thousands of different devices, with those patients moving from location to location on a regular basis. All those devices have to connect to one another and to a central system, with 100 percent uptime, complete data security, and real-time reporting.

Not that easy.

Smarter Standards

The first step in creating smarter hospitals is to establish standards that will enable different types of devices, from different manufacturers, to communicate with one another. That's being done with the Integrated Clinical Environment (ICE) standard, which defines the necessary control, data logging, and supervisory functionality to create intelligent, connected healthcare systems.

 Note

> Technically, the ICE standard is identified as ASTM F2761-2009. The ASTM International (formerly known as the American Society for Testing and Materials) is a nonprofit organization that develops and publishes voluntary technical standards in a variety of industries. The ICE project is currently under the auspices of ASTM subcommittee F29.21, "Devices in the Integrated Clinical Environment."

Figure 11.9 shows the functional elements of the ICE standard. It all starts (at the bottom of the diagram) with the patient and ends (at the top) with the clinician. In between are layers of medical devices, interfaces, network controllers, and an ICE supervisor. It's all a way to get information about the patient—and that affects the patient—to the clinician.

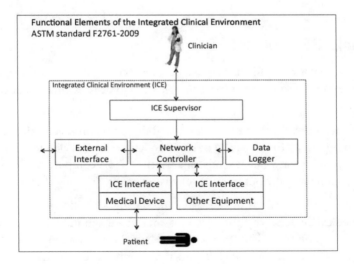

Figure 11.9 *The functional elements of the Integrated Clinical Environment standard.*

The key elements in the ICE standard include:

- **Network controller** that generates alarms if a connected device malfunctions and also provides historical data logs.
- **Network supervisor**, the programming that provides the intelligence in the system and includes clinical decision support, smart alarms, and record-keeping functions.
- **Network interface** that connects medical devices with the network controller.

In the ICE environment, all devices—from BP cuffs to intravenous pumps—are interconnected using plug-and-play technology. The connections and communications between devices are standardized to ensure interoperability. The goal is to ensure the safe integration of medical devices from various manufacturers—which does not exist in today's world—where each manufacturer employs its own proprietary technologies that don't communicate well with those from other manufacturers.

The ICE standard is relatively new. Cross-disciplinary work on it began in 2006, and the initial standard was issued in 2009. The healthcare industry is large and ponderous, and doesn't change easily or rapidly, so it will be some time before these standards see widespread adoption. Still, ICE represents a roadmap to healthcare's connected future and the ultimate evolution into the Internet of Things for Medical Devices.

Other Smart Equipment

The implementation of ICE is just one necessary component of the hospital of the future. We'll also see new smart equipment and devices enter the medical workplace.

For example, the folks at 3M have developed the Littmann Electronic Stethoscope, shown in Figure 11.10. This nifty little gizmo lets your physician focus on listening to your heartbeat while it sends data wirelessly to a computer or mobile device. This one sounds so obvious that it's surprising nobody's thought of it before.

If you're in the hospital for some reason, interfacing with all these increasingly smart medical devices, why not sleep on a smart bed? I'm not talking about one of those "sleep number" beds that lets you adjust the mattress firmness, but rather a holistic system with embedded sensors that monitor the patient's vitals—and more—and feed them back to the nurse's station.

Figure 11.10   *3M's Littmann Electronic Stethoscope, with its digital readout.*

For example, BAM Labs offers just such a system, called the Smart Bed Technology Solution. It employs a special sensor mat, shown in Figure 11.11, that is placed underneath the normal mattress. This mat monitors heart rate, respiration rate, motion (the patient's changes in position), and presence (when they exit and enter the bed).

This biometric data is transmitted to the Smart Bed cloud platform (because all data must eventually reside in the cloud), then packaged into user-friendly applications that can be viewed on any Internet-connected device. This enables hospital staff and physicians to receive the data and any emergency alerts on their own computers or mobile devices.

If you really want to go high tech, consider the robot doctor, such as the RP-VITA Remote Presence Robot, shown in Figure 11.12, from iRobot and InTouch Health. This little fella enables doctors to conduct real-time clinical consultations when they can't be there in person. The clinician controls the robot with an iPad app and talks to patients through the robot's video-conferencing screen. It's a great way to make more efficient use of a doctor's time and to reach patients in remote or underserviced areas.

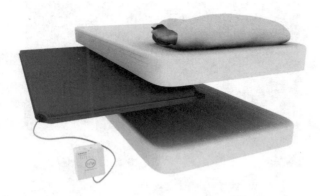

Figure 11.11   *The sensor mat in BAM Labs' Smart Bed Technology Solution.*

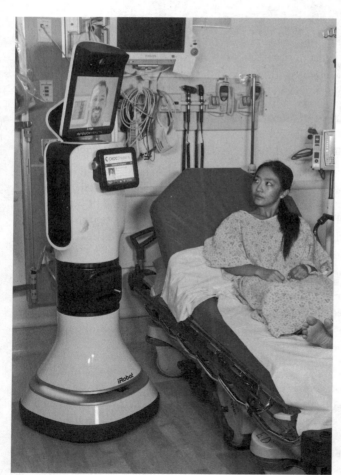

Figure 11.12   *The RP-VITA Remote Presence Robot, from iRobot/ InTouch Health.*

Smart Medical Records

One of the major challenges in today's healthcare system involves your patient records. Each doctor you see keeps his or her own records, as does every hospital and clinic. There's no central repository for your records, no place where everything is stored and collated.

Today, if you visit a new doctor or clinic, they have to call your old doctor's office and request that your paper records be faxed over for them to use. This is time-consuming and expensive, and can delay diagnosis and treatment—and can be life-threatening in medical emergencies when time is of the essence.

It gets worse when you consider all the data gathered by the new class of wearable medical devices. Why shouldn't the data generated by your fitness band, smartwatch, or wearable BP monitor be consolidated with the rest of your medical records?

Of course, all your medical records should be centralized so that every healthcare provider you visit can see everything that's happened to you—and that other clinicians have commented on—in the past. The fact that this hasn't happened yet is because medical records have traditionally been paper records, and they're not easily shared or consolidated.

The key to more effective and efficient medical recordkeeping is to go all electronic. This means keeping new records digitally, in computer databases that can be accessed from multiple providers. It also means digitizing your existing paper records, which is an enormous undertaking.

Once your medical records are all electronic, they have to be stored someplace. Since you need your records to be accessed by all your healthcare providers (and yourself, of course), that means cloud-based storage. When your records are hosted in the cloud, any medical professional (with your permission) can access your records to view the most current information in real time. That should reduce paperwork, speed up processing, and provide faster and more accurate diagnoses.

This type of centralized electronic health record is called a *personal health record* (PHR), and there are lots of companies working on them.

First, many individual clinics and hospitals are creating their own PHRs for their patients. While these services make it easier for people to view their own records at that facility, they typically don't tie into records kept by other facilities and providers.

Beyond that, several big companies and organizations are trying to create PHR databases and applications that span multiple facilities and providers. Some of these are targeted at a given company's employees or the customers of a particular health insurance company. Others are more universal in scope, vying to become an industry standard. In any case, there are a lot of big players trying to get a piece of the pie, as you'll see.

Apple HealthKit

The company that provides your iPhone and iPad (and sells you lots of digital music and videos) would like to manage your health records, as well. Apple's HealthKit service acts as a repository for patient-generated health information and then shares that information with your physician and hospital. The data is collected from the Health app that runs on the company's iPhone and Apple Watch devices (as well as similar apps on other devices) and includes blood pressure, heart rate, weight, and other information.

HealthKit is initially being pitched to help physicians monitor patients with chronic conditions, such as diabetes and hypertension. As of early 2015, Apple has signed up 14 of the 23 top-ranked hospitals in the U.S. for the program.

Dossia

Some of the largest employers in the United States—including AT&T, Intel, and Walmart—have banded together to create the Dossia PHR service (www.dossia.com) for the use of their employees. The Dossia system enables individuals to gather copies of their medical records, in digital form, from multiple sources. They can then create their own personal portable electronic health records and enable access by the healthcare providers of their choice.

The Dossia Health Manager, shown in Figure 11.13, is an app that provides health information personalized for your own family's needs. You see a news feed with all your family's activities and alerts, as well as recommendations to better manage your own personal health. The Health Manager taps into your personal health record, keeping track of medications, allergies, immunizations, doctor's visits, test results, and the like.

FollowMyHealth

FollowMyHealth (www.followmyhealth.com) is a patient portal offered by hundreds of large healthcare providers and facilities. It enables patients to review their medical records, view lab and test results, update their medical information, request prescription refills, communicate with physicians via secure messaging, schedule appointments, and more. Users access the portal via computer, smartphone, or tablet app. (Figure 11.14 shows FollowMyHealth's iPhone app.)

Figure 11.13   *The news feed in the Dossia Health Manager app.*

Figure 11.14   *Using FollowMyHealth's iPhone app.*

MediConnect

MediConnect (www.mymediconnect.net) is a web-based PHR service that assembles most of your data for you. Fill out an online medical record request (and corresponding release form) and a specialist at MediConnect contacts your doctors, pharmacists, clinics, and hospitals to request copies of your medical records on your behalf. MediConnect then digitizes, uploads, and organizes your records to create your MyMediConnect account. There's not much else you need to do.

As you can see in Figure 11.15, a completed MyMediConnect PHR includes all your pertinent medical data, including information about your medical conditions, allergies, procedures and surgeries, vaccinations, medications, doctors, insurance plans, Medicare claims, and family and social history. All this information is in one central location, presented in an easy-to-read and easy-to-understand format. You decide who else can see which information—including your doctors, family members, and such.

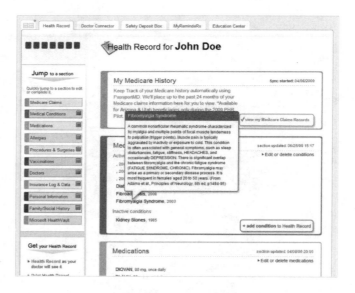

Figure 11.15   *MyMediConnect's Web-based PHR.*

Microsoft HealthVault

Microsoft is betting big on the IoT in general, and on health record services in particular. Microsoft HealthVault (www.healthvault.com), launched in 2007, is a Web-based platform used to store and maintain health and fitness information. It's targeted at both individuals and healthcare professionals.

When an individual creates his or her HealthVault account, the account can be authorized to access records for multiple individuals. Each record contains medical information for a given individual. A single HealthVault account, then, might contain records for all members of a family, enabling parents to manage the records for their children, or older children to manage the records of their senior parents.

Individuals interact with their HealthVault records through the HealthVault website or PC/mobile app that interfaces with the HealthVault platform. (Figure 11.16 shows the Windows 8 version of the HealthVault app.) You can specify types of information to be shared with individuals (such as family members) and healthcare providers; not everyone you share with has to see everything.

Figure 11.16   *The Microsoft HealthVault PHR app for Windows 8 computers.*

You can also upload data collected by various medical devices, using the HealthVault Connection Center. This way your healthcare providers can see your personal fitness information or any other data you collect on your own.

 Note

> Not surprisingly, Google was a big early player in electronic health records with its Google Health service. However, Google Health never gained any traction among consumers or the healthcare industry. It was shut down in 2011.

SMART MEDICINE AND YOU

The future of smart healthcare is upon us. Where we're going is somewhat obvious; how we get there, less so.

It's all about connecting things together—both existing and new things. Currently independent devices will connect together to work smarter and make life easier for healthcare providers and patients alike. Currently isolated records will connect into a single electronic database that both individuals and institutions can access. Everyone will know more about their own health, and be able to act on that knowledge faster and in a more informed fashion.

Some of this connection will happen without any input or knowledge on your part. Hospitals will connect the medical devices in all their rooms, enabling cross-function monitoring and alerts. There's nothing you have to do to make this happen; hospitals are already going down this path.

Some of this connection requires your participation. While it's possible that your doctor or hospital may offer the capability of creating PHRs, you may need to create your own PHR, inputting or uploading your own information, and then sharing with your healthcare providers of choice.

Whomever does the work, the results will be worth it. Whatever medical problems you're experiencing, your doctor will be able to diagnose it faster and more accurately. You and your doctor will be better able to track your long-term health, especially if you have a chronic condition such as high blood pressure or diabetes. If you change doctors or go to the urgent care, all your past results will be immediately accessible, removing guesswork from the process. If you have the misfortune to be hospitalized, your stay will be more pleasant and less prone to clinician error.

In short, the IoT-MD promises smarter, faster, more accurate decisions regarding your medical care. That means you'll live a healthier life—and when something goes wrong, you'll be back on your feet that much sooner.

And let's not forget the cost factor. Yes, it's going to cost real money to connect all those medical devices, build PHR databases, and digitize and organize all those patient records. But that investment pales in comparison to the returns, both personal and financial. The IoT-MD will help stop the upward spiral in healthcare costs and possibly drive costs down. You want lower healthcare and insurance bills? Then hope for a faster adoption of these smart technologies and strategies.

The IoT-MD will save money and save lives. That's smart technology put to good use.

12

Smart Businesses: Better Working Through Technology

The Internet of Things (IoT) promises improved efficiency in offices, factories, and retail stores. In fact, the IoT might make things so efficient that businesses won't need as many workers in the future. Think robots in the warehouse, intelligent assembly in the factory, and tiny chips embedded in office machines that enable office workers to more fully control the devices—or just collect data from it. It's all about automating repetitive and predictable processes, and making all sorts of devices and systems talk to each other.

Experts believe that the IoT for business—what some call the Enterprise Internet of Things, or EIoT—will be the largest of the three main IoT sectors. (The other two sectors are home, which we've already explored, and government.) We're talking 9.1 billion smart devices in businesses worldwide by 2019, according to Business Insider (BI) Intelligence—fully 40 percent of the entire IoT market.

Why are businesses so hot on the Internet of Things? It's partly due to the fact that, of all potential customers, businesses (especially big businesses) have the most money to spend on IoT devices and services. It's also because businesses see the benefits of moving rapidly in this direction.

Smart Offices

Historically, many companies have been quite receptive to new technologies. Think about the technology introduced into the workspace over the past 30 years or so—fax machines, personal computers, laser printers, wired and wireless networks, the Internet, email, instant messaging, tablets, video conferencing... The list goes on and on.

Of course, not all workplaces are as amenable to technological advances; witness the relatively large number of companies still using Windows XP and resistant to the bring your own devices (BYOD) movement. But, all in all, if a company sees productivity benefits, it ends up embracing the new technology.

This is why so many companies are interested in the Internet of Things. The smart office—or intelligent workspace, as some dub it—promises to bring further efficiency and effectiveness to the office environment.

Smart Connectivity

Connected offices are efficient offices.

In the 1950s and 1960s, the only connectivity in the typical office was the telephone. In the 1970s and early 1980s, larger offices were connected to mainframe computers, with "dumb" terminals used for data entry. In the 1980s, offices disconnected from the mainframe in favor of personal computers, and the only connectivity was provided by swapping disks via sneakernet. Connectivity improved in the early 1990s, thanks to local area networks (LANs) and in-house email, and in the later 1990s, thanks to the Internet.

Since the turn of the century, office connectivity has slowly moved away from cables to wireless networking, with more and more workers using notebook PCs instead of desktops. There's also been the reluctant acceptance of portable devices such as tablets and smartphones in the office environment—and the extension of the office to employees on the road or working from home.

The past decade has also seen a boom in virtual collaboration. It used to be that the only collaborations were face-to-face in the company conference room. Today, the proliferation of text messaging, video conferencing, cloud computing, and other technologies make it easy for office workers to communicate with and collaborate with out-of-office employees located anywhere in the world.

Now, all this historical business connectivity practically defines the current Internet—the Internet of people and organizations. What changes will the Internet of Things bring?

As you know, the Internet of Things is about connecting *things*—devices, machines, and the like—rather than connecting people. Since businesses are already adept at connecting their employees, the next step is to link the devices used by their employees, both to each other and to their systems and employees.

Some of this is happening today. Your boss schedules a meeting at 8:00 tomorrow morning using your company's calendar application. This reserves a meeting room and notifies all attendees. The meeting is automatically added to your calendar, which then shows up on your desktop and notebook computers, and your smartphone's calendar app. All well and good, but the morning of the meeting, your boss reschedules the meeting to noon. This change is registered by your calendar app, and your smartphone notifies you of the change—so you can decide to sleep in or do some other work beforehand.

In the future, this process may become more automated. Your boss won't have to manually consult with every attendee; he'll tell his computer or phone that he wants to schedule a meeting and with whom, and the scheduling app will do the rest—"talking" to all the attendees' devices to see when they're free, finding out when the conference room is free, and triangulating all those results to schedule the ideal meeting time. Your boss' phone will also know what he's doing at all times, and when he's getting overworked and running late, and then use that information to reschedule later meetings when necessary.

This is simple activity automation. There's more to come in that vein, including the automation of interviewing/hiring, human resources (HR) management, and your company's internal help desk and external technical support. Intelligent systems will also help to manage employees' work schedules, workspace usage, and the like.

Because more and more employees are working away from the office—either traveling on the road or working from home—this connectivity has to include devices not physically located in the main office. For this to happen, we need always-on connectivity and systems that work across different platforms and devices. And, of course, policies that not just enable this universal connectivity but also encourage it.

The goal, at least partly enabled by the IoT, is to create a virtual connected workspace where actual physical location is irrelevant. Given the proper combination of smart devices, universal connectivity, and virtual systems—including video conferencing, cloud storage and applications, and interactive whiteboards—you should be able to do your job from wherever you and your colleagues happen to be located.

Of course, a lot of this is already happening. Companies around the globe are connecting remote offices (and remote workers) via the Internet and virtual private networks (VPNs). Far-flung workers are collaborating via Google Docs and other cloud-based applications and communicating via Skype, Google Hangouts, and other video chat services. The IoT will only make this phenomenon more commonplace—and more automated.

Smart Environment

On that topic of connected workspaces, not all productivity comes from the activity of office workers. The equipment in your office can connect to the IoT to operate smarter and more efficiently.

Some of this smart connectivity is similar to what's available in the home environment. Smart lighting systems, smart heating/cooling (via smart thermostats), even smart audio systems (playing just the right combination of white noise or ambient music, depending on surrounding noise levels) will make life easier for both office workers and maintenance staff.

Smarter equipment can also make maintenance easier. Consider a trash can (or shredder box) with an embedded sensor that tells the maintenance staff when it's getting full. Maintenance staff can save unnecessary work by not dumping every trash can every day—or alleviate overflows by dumping during the day when necessary.

Then there are smarter printers, copiers, and fax machines. Instead of flashing a light or displaying a message on the front panel when toner gets low or paper runs out, the device itself will notify maintenance staff of the situation. You won't need to call the support guys because there's not enough paper or toner or something's jammed; they'll know automatically.

Or, better yet, they won't know at all because their support robot will have already been sent to take care of the problem. Imagine an R2-D2–like device scuttling down the hallways, refilling printers and copiers, taking care of simple problems and maintenance. You won't need as much support staff—and the basic care and feeding of your office machines will be done more efficiently and effectively.

Consider also the topic of workplace security. Today, you probably have a magic card you wave at or slide through a reader to enter the office or specific areas. With a smarter smartphone (and the necessary app), your phone can serve as your corporate badge, letting you into to whatever areas of the building the company allows you access to. Different employees can be programmed with different restrictions, so that only designated employees get into sensitive areas.

Smart devices can also personalize your environment as you move around the workplace. Your smartphone will contain your preferences for temperature and lighting; using near field communication (NFC) or Bluetooth technology, the parameters of any room you enter will be automatically adjusted accordingly. And when you sit down in a conference room, the room will know you're there and make available whatever apps and devices you need. That might mean lowering a projection screen, firing up a projector, and turning on a nearby microphone.

 Note

> To receive communications from your smart device, the workplace will need to be outfitted with iBeacons or similar sensing devices, as discussed in Chapter 7, "Smart Shopping: They Know What You Want Before You Know You Want It."

One system in use today is Robin (www.robinpowered.com), shown in Figure 12.1, which uses Bluetooth beacons to locate employees via their smartphones. Robin automates room booking and can locate employees anywhere in the workspace. It also analyzes conference room usage, so companies can know how their spaces are being utilized—or how much time is being wasted in unproductive meetings.

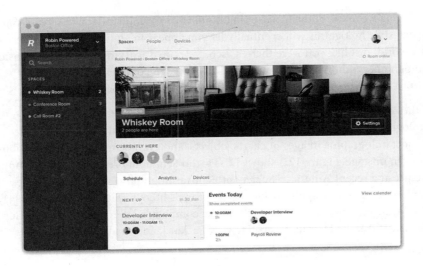

Figure 12.1   *Smart workplace managed from the Robin dashboard.*

Virtual Meetings

Speaking of conference rooms and meetings, who says everybody needs to be in the same room to meet anyway? Video conferencing is already a staple at many companies, large and small, along with interactive whiteboards and cloud-based presentations. And there's more to come.

Instead of remote attendees appearing as thumbnail images along the side of a main display, how about having people attend virtually via a telepresence robot? The Ava 500 Video Collaboration Robot from iRobot (the folks who brought you the Roomba robotic vacuum cleaner) enables employees to establish a remote presence when they're out of the office. Ava 500 looks a little like a kiosk with a flatscreen display on top, on which your talking head appears. There's a camera and microphone mounted above the screen, so you can see whatever Ava sees.

 Note

Ava 500 is the electronic sibling of the RP-VITA robotic doctor, discussed in Chapter 11, "Smart Medicine: We Have the Technology…"

Even better, Ava is mobile. It has wheels so it can travel the office halls and interact with live coworkers wherever they may be in the office. You're not limited to inter-acting with people in the conference room; you can have Ava roam the halls and talk to your coworkers anywhere—even if they're on the move (see Figure 12.2).

If having a robotic doppelganger isn't quite your style, just wait a few years. *Star Wars*–style holograms will eventually let you appear in any meeting, anywhere in the world, just like you were there. Cisco, for one, is working on three-dimensional holographic technology to replace traditional video conferencing in the near future. Just be careful not to put your hand through your boss' face!

Other companies are also jumping into holograms feet first. Trade Show Holograms, for example, markets a hologram projector that enables clients to project holographic images of charts, objects, and even people into their exhibits or even onstage. Figure 12.3 shows Dr. Partho Sengupta presenting a holographic lecture at a recent American Society of Echocardiography (ASE) Scientific Sessions conference in Minneapolis, using the company's hologram technology. This is real stuff available now, not science fiction; the company's other clients include General Electric (GE), the Detroit Tigers, Hewlett Packard, and the U.S. Army.

Figure 12.2   *Carrying on a virtual meeting with the Ava 500 Video Collaboration Robot.*

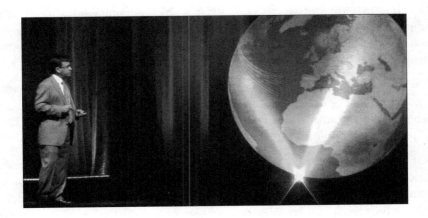

Figure 12.3   *Dr. Partho Sengupta presenting a holographic globe, using technology from Trade Show Holograms.*

Smart Stores

Retailers are a particular breed of business that look to benefit significantly from the Internet of Things. Businesses will use smart technology to better interact with their customers and to better manage the products they sell.

As to the customer interaction end of things, we discussed how the IoT will affect your friendly neighborhood retail store back in Chapter 7, so there's no need to repeat that information here. Instead, let's look at other ways the IoT will impact bricks-and-mortar and online retailers.

For retailers, the most obvious impact of the IoT is on inventory management. We'll talk more about inventory management in general later in this chapter, but know that it's very important that retailers have the right products on their shelves at the right time. Running out of a fast-selling item is just as costly as having too much of a slow-selling one.

Placing smart radio-frequency identification (RFID) sensors on store shelves, or embedded in the products themselves, will help retailers know where all their stock is at any given point in time. Combined with smart inventory management systems, retailers will know when an item is about to run out and should be replenished from the stockroom or warehouse, or even reordered from the manufacturer.

Walmart, for example, has been at the forefront of intelligent inventory management. When an item is scanned at checkout, the inventory level for that item is automatically updated. This helps the company determine when it's time to restock their inventory.

As to how that inventory is restocked, today it's by minimum wage workers. In the future, restocking robots will probably do the job, as instructed by the intelligent inventory management system. Fewer workers means lower costs and more efficient restocking.

Smart item controls will also help a retailer determine where in the store specific products should be displayed. For example, Heineken is working with Walmart to determine where shoppers are going within the store to purchase their beer. This helps not only in product placement, but also in forecasting sales.

Not surprisingly, sales forecasting is another huge IoT benefit for retailers—and for consumers. It's more than just rapid restocking; it's analyzing sales trends and knowing when to order more products to meet anticipated customer demand. This will not only result in more sales, but also keep overstocked or unsellable merchandise to a minimum.

 Note

Intel estimates that retailers' lost sales from out-of-stock merchandise (and deep discounts to move overstocked products) cost retailers $818 billion each year. Ouch!

For those retailers who deliver their products to consumers, the IoT promises smarter and more efficient deliveries. Building on existing global positioning system (GPS) technology, the IoT will better route delivery vehicles to their destinations—avoiding traffic slowdowns, accommodating perishable items, knowing when customers are likely to be home. This will not only get purchases to customers faster, but also reduce fuel and maintenance expenses.

How do retailers handle broken or malfunctioning merchandise? Today, it's the crowded and typically understaffed Returns desk. Add the IoT to the equation and retailers (and manufacturers) can track returns and post-sale service on a real-time basis. This will help them identify trends and warranty issues—and ultimately improve the products they sell. Retailers will know if there's a developing problem with a particular product and allocate resources (including having replacement parts or products available) accordingly.

In short, we're talking smarter, more efficient retail management that better serves customers while reducing the business' costs. Everybody wins (expect maybe the minimum wage worker replaced by robotic restockers).

Smart Inventory Management

Now to the issue of inventory management, important not just for retailers but for every entity along the supply chain. We're talking manufacturers, distributors, warehouses, delivery firms, and the like. How can the IoT help put the right merchandise in the right place at the right time?

Some experts believe that when it comes to the IoT in business, the manufacturing, transportation, warehousing, and information sectors will account for the largest piece of the pie. *BI Intelligence*, for one, estimates that total IoT investment in these sectors will reach $140 billion over the next five years. That's a lot of money that will be spent—and for good reason. The IoT will significantly impact how the entire supply chain operates.

It's all about connecting all the pieces and parts of the supply chain operation, as well as tracking (to the second) every individual item involved. Then it requires intelligent systems to make sense of the collected data and make informed decisions about manufacturing, warehousing, and transporting those items.

These intelligent systems will result in real benefit to all parties involved. We're talking about:

- **Reducing asset loss**, by discovering product or supply issues in time to find and effect solutions.
- **Reducing transportation costs**, by optimizing fleet routes to reduce "deadhead" miles, as well as by optimizing product deliveries.

- **Reducing out-of-stock items**, by projecting future needs and employing just-in-time (JIT) inventory management throughout the supply chain.

- **Reducing overstocked items**, also by projecting future sales and using JIT inventory management and manufacturing.

- **Ensuring product stability**, especially with perishable items, by monitoring the "cold chain" and reducing the one-third of food items that perish in transit each year.

- **Gaining customer insight**, by embedding sensors into all products and analyzing purchase (and possibly post-purchase) behavior.

Most of these benefits come from the use of RFID technology attached to or embedded in individual products. As we've previously discussed, RFID tags broadcast short-range radio signals that can then be tracked by monitoring devices. In the retail environment, RFID tags are read during checkout. In the manufacturing and warehouse environments, RFID tags can be read anywhere to keep track of each item's physical location.

Smart Manufacturing

Smart inventory management starts on the factory floor. It's all about making the manufacturing process more efficient—better quality products built faster, and in time, to meet anticipated demand.

Automation is not new in the manufacturing environment. That automation has been rather large and stationary, however—big machines connected to controllers that control other big machines.

What is new is the deployment of more and smaller sensors throughout the manufacturing process. Instead of relying on a few fixed inspection points, smaller sensors are placed more frequently on the line. Many factories are also attaching sensors to their human inspectors (via smart vests and similar clothing), turning them into living, walking sensors. This enables the manufacturing process to be monitored from any point; purpose-built electronic sensors can spot problems and delays faster than the human workers can.

 Note

Sensors worn in worker vests can also help prevent accidents, track a worker's temperature and other vital signs, or monitor for dangerous gasses or chemicals in the workplace.

Individual components can be monitored throughout the process via embedded RFID tags. RFID tags are also built into the manufacturing equipment, so that their operation and output can be monitored. Managers will know when a particular piece of equipment isn't performing up to specs, or when it needs to be recalibrated or repaired.

Human workers aside, the factory floor becomes more efficient by having all the various machines talk to one another without the need for human interaction. It's a matter of machine-to-machine (M2M) communication. When one machine gets a little behind, it communicates that to other machines up and down the line, which adjust their workflow accordingly. If a machine breaks down, that information is automatically sent to other machines, and the manufacturing process is either rerouted or paused until the first machine goes back online. Machines will automatically notify human workers (or robotic inventory managers) when they need more raw materials, and the appropriate worker or robot will deliver those materials in a JIT fashion so that the line never stops or slows down.

The goal is to create a "factory of the future" where manufacturing evolves from a patchwork collection of isolated silos, each working on discrete parts of the final product, to an environment where all processes are fully integrated into a seamless whole. Thus connected, the entire process can be more easily monitored from a central dashboard, where employees can access both historical and real-time performance. The manufacturing process gets quicker, due to decreased downtime (fewer machines down, less waiting to restock raw materials), and yields increase. Everything just works smoother, with less human interaction needed.

Here's an example. GE's Durathon battery factory in Schenectady, New York, has tens of thousands of sensors installed on its assembly line, and even more tiny sensors embedded in every battery it produces. This creates a massive amount of real-time data that managers use to more effectively manage the entire production process. The sensors gather all sorts of environmental and manufacturing data— the precise temperature and humidity on the factory floor at any given time on any given day, for example. This data can then be correlated with other data to discover how temperature and humidity affect production levels or quality of the finished product.

When the data is properly analyzed and the results applied, managers can adjust those environmental parameters to maximize results. Over time, the factory realizes less waste, higher output levels, and better quality products.

The improvements don't have to be dramatic to have significant impact. A 5 or 10 percent improvement in yield or similar reduction in product defects can return millions of dollars for a tech-savvy manufacturer.

Smart Transportation

Smarter manufacturing processes also impact the next step in the supply chain—transportation. The more precision involved in predicting manufacturing end dates, the more efficient the resulting transportation of the final product from the factory to distributors and retailers.

Of course, smart transportation isn't just for moving products out of the factory. It's also necessary for moving raw materials into the factory, at just the right time to avoid materials-based slowdowns. The more precisely a company can project when raw materials are needed, and the more confident it can be of the shipment of those materials, the more precisely it can schedule incoming deliveries.

Smart scheduling is just part of the transportation equation. Better tracking (enabled by sensors embedded in pallets, boxes, and individual products) will make for more efficient handling and transfer of products between suppliers and the manufacturer, and between the factory and distribution centers and retailers. The manufacturer will know at any given moment where each shipment is, and where it will be next and when.

It's a matter of improving the visibility of products at all stages of the transportation process—and being aware of environmental and other conditions along the route. Smarter tracking results when you know the driver's average speed, how many hours of sleep he's had, traffic conditions en route, current and predicted weather (rain and snow slow things down, naturally), and the like. The more variables known, the more accurately you can predict when a shipment will arrive at its designated destination.

Let's say that your plant has just shipped out a truck full of the most in-demand holiday items from your plant in St. Louis, Illinois, to a big retailer in the Chicagoland area. With the history of the vehicle and driver in hand, weather forecasts available (there's light snow expected in Peoria) and local traffic being constantly monitored (there's a slowdown in Springfield), systems can accurately predict exactly when the truck will arrive at the Chicago store. The store can then have its loading dock staffed and ready to receive the shipment and the in-store workers prepped to put those hot little items on the shelves. Not a minute is wasted.

In addition, by tracking each individual pallet or product, loss can be minimized. If a shipment happens to arrive a couple of boxes short, you'll know where those boxes went because you can track them individually. Theft is minimized, "spillage" is reduced, and you get more of your product where you want it.

Smart Warehousing

The tracking of individual product items continues in the warehouse—whether that's the manufacturer's own distribution center, a wholesaler's warehouse, or a retailer's warehouse or on-location storage area. You (and the distributor and retailer) will know where each item is, up to and including when it's purchased by the final customer.

Modern warehouses are surprisingly high-tech and sophisticated operations. A typical warehouse holds thousands if not millions of SKUs (that's *stock keeping units*, the industry shorthand for each individual product), each stored in a specific location (aisle and shelf). When an order comes in, it's likely to include multiple SKUs, requiring someone to "pick" those items from around the entire storage facility.

How do workers know where each SKU is stored? In the old days, they relied on paper "maps" of inventory locations. Today, the location of each item is stored in a massive database, which is referenced when it's time to fill an order. It works something like this.

A customer places an order for a half-dozen different items. That generates a picking ticket in the warehouse, with each item on its own line. Next to each item is the warehouse location for that SKU. Smarter systems will organize the items ordered for faster picking, so that workers don't have to travel back and forth across the building—they can move in the shortest distance possible to pick each succeeding item.

Even smarter systems automate the process even further, using mobile fulfillment robots. Amazon, for example, has deployed more than 15,000 mobile robots from Kiva Systems to scoot across the floors of its 10 biggest warehouses, programmed to find those items necessary to fill customer orders. (Figure 12.4 shows Amazon's fleet of Kiva robots.)

 Note

Kiva Systems is a subsidiary of Amazon and supplies fulfillment robots to its parent company and other large distributors and retailers, including Walgreens, The Gap, Office Depot, Crate & Barrel, and Saks 5th Avenue.

Amazon's robots (actually, the inventory system that drives the robots) know where each one of the millions of SKUs resides, calculate the shortest distance to travel, and then navigate there. The robots navigate by scanning codes located on the warehouse floor, following commands beamed wirelessly from the master control system.

Figure 12.4 *A few of the Kiva robots Amazon utilizes in its warehouse fulfillment process.*

The little robots slide under the shelves and carry them to a central location, where human workers pick and box the products. Employees no longer have to walk the floors; staying in one place is less physically draining, while having the robots move the products to the people speeds up the entire process—up to three times faster than before. The robots also reduce the warehouse's operating costs by 20 percent, which ain't small change.

The Kiva robots, however, do not actually pick products off the shelves; that's still left to human employees who travel with the robots throughout the warehouse. Kiva and other companies are working on robots that can grasp items, but that's still a few years away. When grasping robots become available and affordable, expect to see further gains in warehouse productivity—and fewer human employees on the payroll.

Smart Management

In the battlefield of the marketplace, smart businesses have a secret weapon, and it's called *information*. The more information a business has, the smarter decisions it can make.

In the Internet of Things, all the data collected at all phases of the manufacturing/ transportation/warehousing process feeds back to a central management system. This data can then be analyzed for trends and other factors, to help company management better plan future operations.

Take the warehouse, for example. In the old days, items in a warehouse might be organized by part number or size. With the information gathered by today's smart systems, stock can be positioned where it makes for the most efficient picking and boxing.

Consider the information available to management. You know not only which are your best-selling items, but also which items customers tend to purchase together and when. This way you can organize SKUs that are typically ordered together in the same area of the warehouse. When purchasing patterns change by season, you know to reorganize the shelves accordingly. You also know what sizes of boxes to stock and when, to maximize the packing part of the operation. Some systems can even predict how many of which items will be sold on a given day, so that some picking and packing can be done *before* the orders are placed!

The IoT will collect more data from more points in the process, and that will make for better management. For large businesses, this means more efficient manufacturing, less waste, improved product quality, faster shipments, and fewer out-of-stocks. JIT manufacturing and inventory management mean keeping less stock on hand, which ties up less capital—and reduces the risk of unsold inventory if sales unexpectedly slow down (or never materialize).

Smart devices + smart processes = smart management. It's a winning equation.

SMART BUSINESSES AND YOU

What does all this smart business stuff mean to you? It depends on where you fit in the grand scheme of things.

If you're an office worker, the IoT will make your office smarter. You won't have as many missed meetings or garbled communications. You'll spend less time waiting on this, that, or the other thing, and be more productive. And the workplace will be more comfortable for all.

Add cloud computing into the mix and you can realize these productivity benefits even if you never set foot in the main office. When it's time for a meeting, you'll attend via video conferencing or holographic representation. You'll collaborate on projects via cloud apps with colleagues spread far and wide. Your time will be better managed, whether you're working from your home office, local coffeehouse, or car.

The company operations under your control will also be more efficient. If you're in charge of manufacturing or distribution or whatever, you won't have to worry about products going out of stock—or having too many unsold items lying around. Everything gets more automated, and you're in better control of things. You will literally know where every single item you make is at any given time, and then be able to act on the information accordingly.

Not that you'll need to, of course. That's because automated management systems will do most of the decision-making for you. You may need to step in if something out of the ordinary arises, but absent that, your job gets easier.

Of course, if your job gets too much easier, you have to question whether or not the company still needs you. The automated systems that are part and parcel of the IoT are going to make a lot of human management redundant. Hell, the automated systems will probably do a better job of things than you ever did, monkey boy. If you're toiling away in this type of middle manage-ment position, better start dusting off your resume.

Same thing goes if you're one of the blue collar workers employed in the fac-tory or distribution center. Just as machines have automated many factory workers out of previously well-paying positions, newer automated processes will get rid of those few humans who are left. Smart devices and sensors will do a much better job of monitoring processes than human inspectors ever did. Management's goal in all this isn't to increase employment rolls, after all.

This will be most apparent in the distribution center, where next-generation robots will be able to pick individual items off the shelf, pack them gently in the proper boxes, add just the right amount of packing materials, stick a label on the box, and hand it all over to the self-driving delivery van or drone. This sort of manual labor is going to go away, there is little doubt about that. It may take a year or two or five, but before long, we won't need human beings in the warehouse.

All this coming automation looks great for big business, but less so for the little people who've been providing the labor all these years. Amazon's and Walmart's fulfillment costs will go down, which might translate into lower prices for consumers. But there will be fewer consumers with money to buy those products, because all the manual laborers will have been replaced by robots and automated systems.

And this, dear friend, is the price of progress.

13

Smart Cities: Everyone's Connected

If everyone, every home, and every business is connected via the Internet of Things (IoT), why not take the next step and connect your entire neighborhood, or even the whole city? Connected devices can help reduce congestion on local roadways, alert the fire department in case of emergencies, and even signal the need for road maintenance or additional police patrols. The smart city might mean a smaller government—or a more intrusive one. It all depends.

Understanding the Smart City

The Internet of Things has a lot to offer municipalities of all sizes. It's not surprising, then, that there's a lot of talk—and a bit of action—on the whole "smart city" front. In fact, there's a recognized smart city concept floating around, kind of a template or set of guidelines that cities across the United States and around the world can model going forward.

The goals of the smart city are to make better use of public resources, increase the quality of services offered to citizens, and reduce operational costs of the public administrations. To achieve these goals, the smart city must deploy an infrastructure that provides simple and economical access to most if not all public services, including transportation and parking, lighting, utilities, surveillance and maintenance of public areas, and more.

What does this theoretical smart city concept mean in practice? We're talking about all sorts of potential public benefits, including more efficient traffic flow, better management of public buildings and areas, reducing the (huge) costs of public lighting, better managing waste removal and other public utilities, and more effective policing and emergency services. These are all worthwhile goals.

As with all IoT-related activities, the smart city will be powered by data collected from smart devices of all shapes and sizes. All this data can also be used to increase the transparency of the local government, enhance the awareness of the public about the status of their city, and stimulate citizens' participation in the public administration.

As exciting as all this is, the smart city (at least in the United States) is still a ways away. There are the obvious technical and financial challenges, of course (it's all rather complex and expensive to implement), but cities face additional political issues. With literally hundreds of billions of dollars in spending at stake, who makes the purchasing decisions? And who benefits from local and state contracts? Who makes the decisions, especially when technology must necessarily bridge multiple adjacent municipalities? And how do governments deal with the privacy and security issues inherent with this massive level of data collection and de facto public surveillance?

Once these challenges can be overcome (and they inevitably will be), expect your city to install thousands of sensors on local streets, parks, and buildings, the better to gather information about public usage, air quality, noise levels, and the like—and then use the collected data to provide better and more efficient services. The future is bright for cities and citizens everywhere.

Smart Infrastructure

A smart city starts with enhanced smart infrastructure. The IoT is built on the collection of various types of data, and the typical city offers a plethora of data just waiting to be collected. That means building out an infrastructure that includes the necessary sensor devices, of various types, and a communications network to link them all together.

Consider the lowly vehicle bridge, a crucial component of our transportation infrastructure. Bridges across America are aging and need significant maintenance to avoid the sort of catastrophe witnessed in 2007 when the I-35W Mississippi Bridge collapsed in Minneapolis. To properly monitor the continuing safety of a bridge requires a variety of sensor types, including temperature, humidity, vibration, and pressure sensors. In earthquake-prone regions, add seismic activity sensors and accelerometers to the list.

All these sensors have to be connected to a central system that monitors data in real time and analyzes the historical data for trends. If vibration levels go up over time, that means something. If sensors measure an increase in pressure combined with an increase in temperature and humidity, that means something else. All the data is relevant, and all must be collated and analyzed.

Or consider the unique situation of road tunnels, which are key to the transportation flow in many cities. Not only must the condition of the roadway be monitored, so must the condition of the tunnel structure itself. Then there's the need to monitor air quality in the tunnels, which requires sensors to detect air flow, carbon monoxide (CO), carbon dioxide (CO_2), nitrogen dioxide (NO_2), and a variety of other gases. The output from these sensors must be monitored in real time (to trigger alerts when something gets out of whack) but also analyzed in the longer-term to discern trends that might indicate tunnel maintenance is necessary.

That's part and parcel of what the IoT is all about, of course. The challenge is deploying all the necessary sensors and then networking them together in an efficient manner.

A city, then, must construct some sort of low-power wireless sensor network (WSN) to connect these and other sensors in the public purview. We're talking hundreds of thousands of sensors monitoring roadways, streetlights, parking meters, public buildings, parks, and other public areas. These networks must be active 24/7 with 100 percent uptime over a relatively large area and be able to function with minimal human interaction.

While several wireless technologies are usable on a city-wide basis, the one getting the most attention today is 6LoWPAN. This wireless protocol is designed specifically for low-power devices transmitting relatively small amounts of data, such as sensors. It's capable of supporting large numbers of these devices over a metropolitan-wide network. In this aspect, connecting a smart city is totally different than the local area network (LAN) you have in your home; the amount of land to cover and the number of devices to support are major challenges.

 Note

The name 6LoWPAN is an acronym for IPv6 over Low power Wireless Personal Area Networks. Other relevant protocols for city-wide networks include DASH7 and MyriaNed.

As you can see in Figure 13.1, a city-wide 6LoWPAN network is comprised of multiple smaller LoWPAN networks, serving smaller geographic areas. Each of these individual LoWPAN networks connects to the main network server (and to the Internet) via *edge routers*. The edge router routes traffic into and out of the LoWPAN, as well as handling network discovery for devices within that LoWPAN. Smaller nodes are housed within each router, serving as either hosts or routers; individual sensors and devices are connected to these nodes, and thus to the larger network and the Internet.

Whatever wireless protocol is employed, a city's smart network will also need to connect to those relevant IoT sensors and devices installed in private homes and businesses. We're talking important emergency sensors, such as smoke alarms and security systems, as well as sensors that monitor energy usage and building temperature.

At the head end of the network, then, a city must install operating software capable of monitoring, analyzing, and acting on this mass of collected data. It's virtually impossible to leave the monitoring and analysis to human employees; cities have to rely on smart systems to do the watching and (most of) the thinking for them.

Smart Communication and Emergency Management

A city has a lot of public employees, well beyond the staff in the city offices. We're talking (for many cities) police officers, firefighters, emergency responders of all sorts, garbage collectors, snow plow drivers, road workers, and other maintenance workers. The city needs to communicate quickly and efficiently with all these employees, especially in emergency situations.

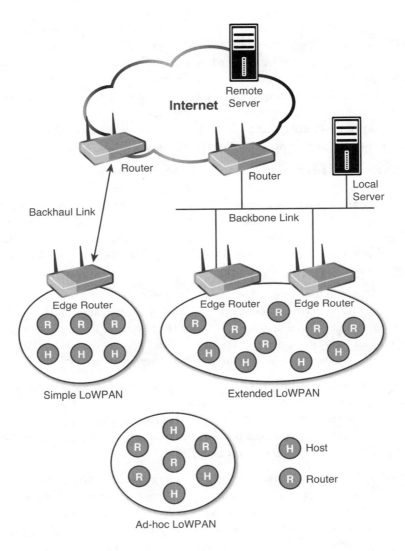

Figure 13.1   *The architecture of a city-wide 6LoWPAN network.*

Consider the all-too-common spring scenario where a rash of potholes needs to be repaired. Today, this occurs (if it happens at all; some municipalities are better at this than others) when someone, typically a local resident, notices the pothole and calls into City Hall complaining about it. The city routes the complaint to the streets department, who puts it on their list. Eventually, a crew is assigned to fill the pothole, typically by receiving a printed list of issues to address on a given day.

Now see how this might work with the IoT, with the appropriate embedded sensors networked into smart systems. Instead of waiting for a citizen to complain about a pothole (typically because he lost a hubcap while driving into the thing), smart sensors detect the presence of the hole and report it—complete with its location and size—to the central system. The system automatically enters the issue into its main database and assigns repair to an appropriate street crew, prioritized based on location and severity. Thanks to the IoT, the city knows about the problem sooner and can get it fixed faster. Everybody wins.

Police and other emergency responders also benefit from faster, smarter communications. Security systems are connected directly into the city's central system so that police are notified the instant an alarm goes off. In addition to receiving the alarm, the police also receive information from other sensors in the home or office building—door sensors, window sensors, motion sensors, even security cameras. The police get notice of the problem faster, along with important details to plan their response.

Same things with firefighters. A smoke alarm goes off, the data gets sent directly to the nearest fire department. Before they pull the fire truck out of the station, they know where in the building the fire is active, the best route into and through the building, and what obstacles they might face. They also get the latest floor plans of the burning building, previously stored in their central database.

Emergency medical technicians (EMTs) also get smarter and more effective thanks to the IoT. When a call comes in, they receive data from any wearable medical devices on the citizen, as well as the person's complete medical history. The responders know what to expect before they ever get there and can thus be better prepared for that particular medical emergency.

Consider, also, the implications of the IoT on larger-scale public emergencies. Whether it's a flood, tornado, snowstorm, or hurricane—or mass shooting, explosion, or terrorist attack—authorities will be better informed beforehand and experience more efficient communications in responding to the event. The left hand will always know what the right hand is doing, and be able to better coordinate their response.

Smart Roads and Traffic Management

Cities, whether urban or suburban, have to deal with a lot of motorized vehicles. We're talking traffic control, road maintenance, parking management, you name it. It's a major headache.

The IoT, however, will relieve that headache. How? By using a combination of smart roads and sensor devices, along with intelligent management systems.

Smart Parking

Let's start with the parking problem. If you live in a city of any decent size, you know how hard it is to find a parking place, especially during popular events. Wouldn't it be great if you knew exactly where the nearest open space was, so you wouldn't spend half your time driving around looking for it?

Smart parking technology is out to solve this particular problem. Streetline is a company that makes special parking sensors that cities are embedding in on-street parking spaces. These sensors detect if a car is parked there or not and send that data to a central service. Drivers use the company's mobile app, shown in Figure 13.2, to find the nearest unoccupied spaces. In the future, this capability can be built into cars so that you can have your car navigate to the best available parking with the press of a button or voice command.

Figure 13.2   *Finding the nearest parking space with the Streetline app.*

 Note

Streetline also provides data to cities about usage patterns so that they can revise parking policies, rates, and regulations—and know where more parking spaces are needed. It also manages parking payments so you don't have to manually feed the meter.

Smart Traffic Management

Parking is just part of the urban traffic problem. Traffic itself can be a nightmare, with massive congestion commonplace in many metro areas.

When you want to avoid traffic jams and minimize the number of red lights you have to stop at, it's time to turn to the IoT. By constantly monitoring traffic flow with its roadside sensors, the IoT's smart systems can manipulate traffic signals and even lane availability to make sure the greatest number of drivers get to their destinations with the least number of interruptions.

A lot of the necessary technology exists today. Most cities already utilize some in-pavement traffic detectors, camera monitors, and timed traffic lights. But today's crude mechanical sensors pale in comparison to what tomorrow's smart sensors and devices will do.

Part of what we'll get with the IoT is more sensors monitoring more critical points, providing a denser grid of data. Sensors can be embedded in roadways or installed street-side in traffic signals or light poles; different types of sensors will monitor different things, such as vehicle traffic, air quality, noise levels, and the like. Data can also be co-opted from drivers' smartphones or smart cars.

How does all this work? Quite well, thank you. Carnegie Mellon University recently ran a pilot project in Pittsburgh, Pennsylvania, using a variety of "smart signal" technologies. In this test, motion sensors were tied to traffic signals, which themselves were connected via signal-to-signal communications. The results? A 40 percent reduction in the time spent stopped at traffic lights and a corresponding 26 percent decrease in travel times. (That also translates into a 21 percent reduction in exhaust emissions, another good thing.)

It's simple, really. Manage traffic signals and other activities to keep cars moving as rapidly as legally possible and lots of good things result—drivers get where they're going faster, there are fewer accidents, road deterioration is minimized, and the air quality improves. It's all about Smart Transportation Systems (STS).

Smart Roads

Even smarter traffic management will come when we make the roads themselves smarter. Smart cars are a future that some are looking forward to, but smart roads are closer on the horizon and could have a bigger immediate impact. Let's face it; it's easier to install a few sensors in a local roadway than it is to convert America's entire fleet of automobiles into self-driving vehicles.

Note

Learn more about smart cars in Chapter 8, "Smart Cars: Connecting on the Road."

What exactly is a smart road? It's a combination of several different technologies, all designed to better manage traffic flow and avoid congestion and accidents.

These technologies include the following:

- **Glow-in-the-dark road markings**—These are stripes and other markings made from paint that "charge up" during daylight hours. The markings then glow green for up to ten hours during the night, as shown in Figure 13.3. In addition, temperature-sensitive paint could display glowing snowflakes on the roadway in adverse winter conditions. The result is safer driving due to more visible road stripes.

Figure 13.3 *Glow-in-the-dark road stripes on Highway N329 in the Netherlands.*

- **Smart road lights**—Lighted roadways are safer than dark stretches of highway, but it costs money to power all those lights. Smart roadway lighting uses motion-sensing technology to tell when a car is approaching and then lights that section of highway. The lights grow brighter when a car comes closer and slowly dim as it passes. It's great technology for less-travelled roadways where lighting is valuable but not currently economically feasible.

- **Wind-powered lighting**—If paying the electric bill is an issue, why not employ alternative forms of energy? Wind-powered lights use roadside pinwheels to generate electricity, using wind drafts from passing cars. The pinwheels only revolve when cars speed by, thus lighting up the road ahead for them. In a way, it's self-powered lighting for and by the motorists themselves.

- **Priority lane for electric vehicles**—Speaking of electricity, how about using the highway system to charge up our coming fleet of electric cars? Some experts are proposing the creation of induction priority lanes with embedded magnetic fields, designed to charge electric vehicles while they're on the go.

- **Solar roadways**—Still on the subject of electricity, we can also let the sun provide power for our smart roads. All those millions of miles of roadway are out in the sun; let's harness the solar power for various uses. Several companies are working on glass solar panels that can be embedded into the roadway surface. The energy generated by these solar panels can be transferred to roadway lighting, melt snow and ice, or be used for other purposes, such as...

- **Smart roadway displays**—Instead of relying on the traditional overhead or side-of-road signs, we can use the solar panels embedded in our roadways to display information to drivers via light-emitting diodes (LEDs). In addition to powering conventional road markings, imagine a roadway with lighted arrows alerting you to upcoming lane changes, numbers that display your speed or the posted speed limit, or words and letters that deliver important messages. (Figure 13.4 shows an artist's rendition of just such a roadway.) It's very futuristic but potentially game-changing.

 Note

One of the more prominent companies working on roadway solar panel/ LED displays is Solar Roadways. The company's modular pavement panels, made from 10 percent recycled glass, can withstand 250,000 pounds of pressure, melt snow and ice on contact, and display messages via built-in LED lights.

That's a lot of upcoming smart road technology, in addition to the expected embedded sensors and monitoring devices. The era of the "dumb" road is over; soon, all the surfaces we drive on will exhibit the necessary smarts to make for a safer and more pleasant driving experience.

Figure 13.4   *An artist's rendition of a test Solar Roadways installation.*

Smart Public Lighting

We touched on smart lighting for highways, but smarter lighting options are also coming for all public spaces, indoor and out. It's a matter of making lighting available only when needed—and saving on energy costs when no one's around.

To a large degree, this involves the same smart lighting technology for the home that we discussed in Chapter 5, "Smart Homes: Tomorrowland Today." It's a matter of using various sensors (especially, but not limited to, motion sensors) to determine whether there are people in a given room or public area. If there are, then the smart tech turns on the lights. If there aren't, the tech saves energy by keeping the lights off.

 Note

Cities don't want to get too carried away with smart lighting solutions—you don't want *all* the city's lights out after dark. Lighting, in many cases, is part of a city's identity; lighting also affects residents' sense of security—and, in many instances, enhances that security.

Public lighting is a big expense for most cities—approaching 20 percent of all electricity consumed. If that power consumption can be reduced by even a small amount, the savings can be considerable. (The reduced energy usage is also good for the environment, of course.)

One increasingly popular approach is to switch from traditional incandescent or fluorescent lighting to newer LED lighting technology. This can cut a city's energy costs in half, without turning off a single light. Add smart lighting controls to the mix, and you're looking at up to 80 percent savings from existing levels.

Smart Utilities

Many cities run at least some of their own utilities—water, waste management, even gas and electricity. Keeping costs down for both the city and its residents is an important challenge.

Smart Waste Management

Garbage collection can be a big mess. Not only is the collection costly and time-consuming, you have the issue of where to put all that stuff. (Spoiler alert: It's probably your local landfill.)

When it comes to making waste management a little smarter, let's start with intelligent waste containers. Right now, all your trash goes into a big green (or yellow or blue) trash bin that the garbage truck collects and dumps once a week, whether it's full or not. In the IoT future, imagine a container with embedded sensors that detect how full the bin is and summons the truck only when necessary. This will reduce costs by optimizing the truck's route and make for more efficient collection.

One such intelligent waste bin is the BigBelly, shown in Figure 13.5. This is a solar-powered waste bin and trash compactor that alerts sanitation crews when it needs to be picked up and dumped. It's great for industrial clients; Boston University says that BigBelly has helped it reduce its trash collection from 14 times a week to less than 2 times. That's noticeable.

Smart Water Management

Municipal water management is also a major source for IoT optimization. Right now, most cities still employ meter readers to walk through neighborhoods and take manual meter readings. That's very, *very* old school.

Going forward, cities will install smart water meters with embedded sensors and radio-frequency (RF) transmitters to monitor individual household water consumption. By collecting real-time data, the water utility can notify customers (via phone, email, or text) if usage levels are unusual, thus warning of potential water leaks. This data will also provide cities with insight into overall usage trends on a neighborhood-by-neighborhood basis. (They can also compare water usage with other collected data—temperature and precipitation patterns, for example.)

Figure 13.5 *The BigBelly solar-powered intelligent waste bin.*

Smart Grid

Another public utility that deserves even more attention is the electric company and the overall electric grid. Power usage can get a lot more efficient when we connect various smart devices to a new smart grid.

Understanding the Smart Grid

What is this smart grid we've been talking about throughout the book? Put simply, it's a modernized electrical grid—the collection of power plants, transformers, and transmission lines that bring electricity to all the homes and businesses in a given area.

The electrical grid, smart or otherwise, is what you plug into when you turn on your TV, flip a light switch, or power up your computer. Our current grid was built over a century ago, although it's seen some technology improvements since then. It consists of close to 10,000 electric-generating units connected to more than 300,000 miles of transmission lines. Truly an engineering marvel, capable of generating more than 1 million megawatts of electricity, the grid is starting to show its age. Large pieces of the grid have been patchworked together, and we're stretching its capacity to the limit.

Moving forward, we need a smarter grid, one built from the ground up to handle larger power loads—and better manage those loads. This new grid, the so-called smart grid, will use digital communications technology to collect and disseminate data about energy usage—the behaviors of both consumers and suppliers. That data can then be used to improve energy efficiency, helping consumers use less energy and save money.

The smart grid will enable two-way communication between the utility and its customers, which will let smart home devices talk to the utility and make for more efficient energy usage. This new grid will also contain sensors along transmission lines to better monitor power usage in all possible ways and respond more quickly to changing energy demand.

Smarter Energy Management

This last point means putting power where and when it's needed. More efficient distribution of the power load will result in fewer power outages and brownouts and, as power usage will be better-balanced, require less power overall.

Power usage can also be managed by having the utility communicate with smart devices installed in homes and businesses. We're talking smart thermostats, smart appliances, and the like that can receive instructions from the power company and shift their usage to times with lower demand.

For example, during a hot summer day when energy usage is peaking, the utility might send out instructions to cut power for nonessential operations, such as turning up the temperature for air conditioners and turning off appliances such as dishwashers and laundry equipment. This will not only help alleviate overall demand for electricity, but also lower usage rates.

A Self-Healing Grid

Experts say that a smart grid will also be a more easily maintained grid. Today, a small power outage quickly becomes a larger one, thanks to the domino effect of failures cascading along the line. Companies often don't know where problems lie until customers call in to complain. And it often takes an inordinate amount of time to repair the damage and bring blackened areas back online.

A smart grid will better manage blackouts and damages, due to what the industry calls *distributed intelligence*. This involves capturing data at the "edge" of the grid, where the electricity is consumed, by smart power meters and sensors in homes and businesses. By analyzing this data on the spot in real time, more information and faster decisions can be made, especially in emergency situations.

Distributed intelligence will enable utilities to know about outages at the moment they occur, well before customers call in. The power company will be able to quickly pinpoint the source of an outage so that repair crews can immediately be dispatched to the problem area.

The smart grid will also enable utilities to better isolate power outages before they affect the rest of the grid, rerouting power paths around the problem area to keep the power flowing to the majority of customers. This will harden the entire system against all manner of emergencies, from severe storms to sunspot activity to terrorist attacks.

In addition to minimizing the extent of an outage, the smart grid will also be more easily repairable. Service will be able to come back online more quickly after an emergency, with selective enabling—routing power to emergency services first, for example. In this manner, the smart grid becomes a self-healing distribution system.

 Note

Because it will be built to handle the two-way flow of electricity, the smart grid will also be better able to take advantage of customer-generated power, whether in the form of traditional gasoline-powered generators during an emergency, or solar or wind-powered generators that operate full time. This will help to ease the power load and accommodate the growing movement toward green energy.

Collecting and Using the Data

The power companies, of course, will take full advantage of all the data collected to more effectively manage their resources and infrastructure. Companies will get a better handle on customer energy usage and manage supply to better match this demand.

All the data generated by the smart grid will (or at least should) be available to consumers, too. The more you know about your own energy usage, the smarter you can be about what you use and how. You'll have access to real-time data about energy usage by time of day, and what devices exactly are using all that electricity. You'll also see how much that usage costs, which will help you save money by using less power when electricity is most expensive.

Building the Smart Grid

Creating a smart power grid is going to be a lot of work and cost a lot of money. The new smart grid will consist of millions of individual pieces and parts—power lines, sensors, controls, computer systems, and the like. Not all necessary technologies are widely available today; some are still in development and need extensive testing before being publicly deployed. It will take time for it all to come together.

Consider the following technologies thought to be part and parcel of the smart grid:

- **Integrated two-way communications**, either via traditional wired networks, power-line networks (a natural, if you think about it), or wireless networks. These communications are necessary for real-time control, data collection and exchange, and security.

- **Sensing and measurement**, to monitor electricity flow and equipment status, evaluate congestion and grid stability, and prevent energy theft. All manner of sensors may be deployed, including those to measure voltage and wattage, ambient temperature and humidity, weather conditions, electromagnetic signature, energy leakage, and the like. Included in the types of sensors are...

- **Phasor measurement units (PMUs)**, high-speed sensors distributed throughout the entire transmission network to monitor the state of the system. These sensors take up to 30 measurements per second, representing the magnitude and phase of the alternating voltage at a given point in the network. PMUs enable automated systems to quickly respond to evolving system conditions in a dynamic fashion, thus minimizing downtime and preventing further outages.

- **Distributed power flow control**, via devices that clamp onto transmission lines to control the flow of electricity. This technology provides more consistent, real-time control over how energy is routed within the grid, and to store customer-generated electricity.

- **Smart meters**, installed at customers' homes or businesses. These are digital meters (as opposed to the older mechanical ones) that monitor usage in real time and provide automated transfer of information between the power company and customers' smart devices. Smart meters will also provide utilities with more data about how electricity is being used in individual locations and throughout their service areas.

 Note

Smart metering technology is part of what is often referred to as Advanced Metering Infrastructure (AMI). This is because the meters need to be installed in conjunction with other communications infrastructure to transmit the collected data to the utility. The AMI provides a communication path extending from the power-generation plants to the smart devices installed in customers' homes and businesses, and back again.

- **Smart power generation**, to match electricity production with demand by using generators that can start and stop independently of other units. This is called *load balancing* and can be automated via the sensors, controls, and systems of the smart grid.

- **Intelligent control systems,** capable of constant monitoring, instant diagnosis, and appropriate solutions to grid disruptions or outages. These systems will incorporate distributed intelligent agents, analytical tools and algorithms, and operational applications.

To a large extent, the smart grid is being piecemealed together from these and other technologies. Some experts estimate that it will take at least a decade for all the pieces to come together—but when they do, we'll all see immediate and significant impact.

SMART CITIES AND YOU

How will the smart city affect you, your home, and your workspace? The effects will be subtle but significant.

First off, not much will look different. You'll still receive the same municipal services—the same utilities, the same emergency services, the same police and fire protection. You'll drive on the same roads and through the same intersections. Your daily activities will be, on the surface, not much different from today.

Most of the benefits from smart city technology will happen behind the scenes. Smart lighting will always be on when you pass by, but turn itself off when you're gone. Smart communications will result in faster response times if you have an emergency. The smart grid will result in lower electric bills—and have a positive impact on the environment.

Other benefits will be more noticeable. Smart parking systems will make it easier for you to find—and pay for—urban parking. Smart roads will provide more useful information as you drive. Smart waste management will mean that your trash bin will never overflow, no matter how much refuse you generate.

Most of these benefits will be mirrored by your local government—which, in the long run, also benefits you. Savings resulting from smarter utilities and lighting, as well as from the smart grid, should result in lower tax rates. More automated operations may result in lower staffing levels, which also cut expenses and (presumably) your taxes. More information gathered by all these smart devices should result in smarter decision-making from your local government—or so we can hope.

For many of these benefits to accrue, you'll need to connect your city's smart services to smart devices in your home. All those devices discussed in Chapter 5 get even smarter when connected to the smart grid and other services. You want that Nest thermostat to really strut its stuff? Convince your city or local power company to upgrade to smart grid technology. It's only when all these smart devices and systems, in your home and in your community, get connected together that the IoT begins to realize its true potential.

Of course, all this enhanced connectivity and data collection will also mean that your government will know more about you. Do you want your local constabulary to have blueprints of your house and know when specific rooms are occupied? Information that might prove useful in case of emergency seems invasive in an evolving police state. Same thing with energy usage—how important is it for your power company to know what electronic devices you have in your home and how much power they're using?

Consider this scenario. Various sensors report to a central database that a particular room in your apartment is using an abnormally large amount of electricity, is constantly lit, and has higher than normal temperature and humidity levels. Now, this may be because that room hosts a terrarium or large saltwater fish tank, but authorities could read the same data and conclude that you're growing marijuana there—and stage a raid based on that data.

That would not be a good use of the IoT, but you can see how we can get there. There's a flip side to everything, including smart technology. Think about this when your City Council is debating investing in these sort of smart devices and systems; there's a potential downside that needs to be addressed.

14

Smart World: The Global Internet of Everything

After we get done connecting our appliances, lights, houses, cars, streets, and cities to the Internet of Things (IoT), what do we end up with? The global IoT, of course, where devices are connected to other devices and systems located anywhere in the world. Just as the current Internet has few state or national boundaries, so shall the IoT be unfettered by those little dotted lines that mere mortals like to draw on maps. The devices in your home or car don't care where other devices are located; they'll connect to whatever, wherever is needed.

The global IoT is going to change the entire world, not just your home or city. Some of these changes will be common across all locations; other changes will be more specific to a given region or even continent. For many people in many countries, the IoT is going to be disruptive, hopefully in a good way.

There's a lot the IoT can bring to a lot of people. In addition to smart appliances and self-driving cars, we're talking about potential solutions to many major world problems, bringing better healthcare, more efficient transportation systems, new jobs, and even a cleaner environment (and possibly a way to effectively deal with climate change). Big impact in many places.

Scaling the Internet of Things Globally

Examining the global Internet of Things is a lot like looking at the IoT in your home or neighborhood; much of the potential and many of the issues are the same. Smart appliances and other home devices are pretty much the same in every developed country in the world; the same smart coffeemaker works in Beijing or Helsinki just as it does in Los Angeles or Des Moines. As long as local and personal economies can afford the technology, it can be deployed just about anywhere.

That's a good thing for IoT device manufacturers. While there are obvious cultural differences between countries (self-driving cars have to drive on the left side of the road in England and India, for example), the basic technology remains the same. So India's self-driving cars can use the same technology used in those cars from American manufacturers. There's efficiency in that.

In addition, the systems behind the various components of the IoT are increasingly universal. Best practices in healthcare mean that automated devices and systems in American hospitals can also be deployed in China and Russia. Again, some cultural peculiarities may come into play, but connecting all the medical devices in a hospital together has the same benefits anywhere in the world.

Obviously, large enterprises will invest in IoT technology in their offices and factories worldwide. This will help speed the adoption of the IoT. Some of this will be facilitated by the corresponding growth in cloud computing and data storage; it's becoming increasingly less relevant where data originates when it's stored in and accessed from the cloud. A multinational corporation can act on data collected from smart devices in a Chinese factory and apply it to their marketing plan being devised in New York and their sales offices in Dublin.

Cloud computing and the IoT will also enable different companies to better work with one another. Think manufacturers and their suppliers here. It will now become easier and more cost-effective for a small company in Calgary, for example, to supply critical parts to a large electronics manufacturer located in South Korea. The IoT will enable their systems to work with one another just as if they were in the same building. When the Korean plant needs more of the Canadian part, the volume is accessed and the order placed for the shipment to arrive at the factory just in time—even and especially taking into account the shipping time between the two countries. The information is gathered via IoT devices and communicated via a data storage receptacle in the cloud. Distance doesn't factor into it anymore.

Connecting Cities, States, and Countries

One of the biggest challenges in creating this global IoT is the deceptively simple one of connecting everything together. To enable that factory in Seoul to communicate with the parts supplier in Calgary requires a sophisticated network of connections unlike anything we've seen in the past.

Put simply, we need to connect local IoT networks not just to a nationwide network that spans the United States, but also to similar systems in China, India, or Singapore. Each local IoT network needs to be connected to a global network—and thus to other local networks.

This makes the typical home-based or car-based IoT solution look like child's play. It's one thing to connect all the devices in a car or building to one another. It's another thing completely to connect your car in Wichita to the factory in Dusseldorf.

Consider the healthcare industry. Today, a patient is treated by doctors who work with a given hospital. In the future, you may want to consult specialists located half a world away—and the IoT will make that happen. Going forward, it's not enough to connect all the medical devices in a single building; those devices eventually will need to be connected to similar devices and systems in other hospitals and other countries. All it takes is for one system to talk to the other one.

That challenge cannot and should not be minimized. Scaling the IoT is a technological challenge on the scale of building the original Internet. We're no longer talking building-wide local area networks (LANs) or city-wide wide area networks (WANs); we're now talking about this LAN or WAN connecting to a larger network to communicate with a LAN or WAN in another country, often in another language.

How will this cross-country interconnectivity of multiple local networks be accomplished? One solution is to use the current Internet. It's already in place and it's already relatively location-neutral. For example, you have no problem at all connecting your local computer to websites in Germany and India; the Internet backbone goes virtually everywhere.

This should work, especially when you consider that all or most of the devices connected to the IoT have their own unique Internet Protocol (IP) addresses. Just connect a given device to the Internet (either directly or through another network or set of networks) and it should be visible to other similarly-connected networks and devices.

There are challenges, however. One is the sheer size of the endeavor; we're talking 50 billion or more individual devices (by 2020), and that's a lot of things to connect. This challenge is not insurmountable, but it is also not inconsequential.

In addition, there's the challenge of getting a variety of different network technologies to connect with each other and with the IP-based devices on the Internet. How will your home's ZigBee network connect to your city's 6LoWPAN network, and then to whatever networks are being used on the other end of the connection? Network interoperability is crucial.

This cross-connectivity has to happen—and is happening. Cities are talking (digitally) to cities and to state and national institutions. National networks are talking to other national networks. Cities in one part of the world are talking to cities on other continents. We're getting there.

And it's not just the basic communications. One entity's systems must communicate and work with other systems. Your local power utility must not only connect to and communicate with all the smart devices in all the smart homes in your city, but also with other utilities throughout the state, across the country, and around the world. The communication must go up, down, and sideways, without the traditional sense of hierarchy. If data from your home's smart thermostat can improve the efficiency of a factory in Bangalore, networks and systems need to enable that flow of information. There must be no artificial boundaries.

The Rural Internet of Things

We've discussed at length how the IoT can improve the lifestyles and economies of first-world countries, particularly in urban environments. But the IoT also offers benefit to rural areas, especially in less-developed countries.

Worldwide, nearly one billion people live in areas without access to paved or all-season roads. This makes it difficult if not impossible for them to get food, medication, and other supplies on a timely or regular basis.

Enter the IoT, in the form of unmanned aerial vehicles—drones, in another word. These autonomous flying vehicles can fly *over* poor or nonexisting roads, delivering necessary goods faster and more reliably than trucks, motorcycles, or similar vehicles.

One project exploring these possibilities is the Matternet. This is a project that proposes to use a fleet of electric-powered drone aircraft to transport supplies to remote areas.

The prototype aerial vehicles used in the Matternet project are quadcopters with a range of 10 kilometers, as visualized in Figure 14.1. When fully realized, a single quadcopter is envisioned to carry a load up to 1,000 kilograms, or close to a ton. (First-stage prototypes can carry only 1 or 2 kilogram payloads.) These drones will be used to deliver medicine, supplies, even people.

Figure 14.1   *An artist's rendition of the Matternet drone delivering a package.*

The project promises to bring the modern world into remote and impoverished areas worldwide. This use of drone aircraft to deliver supplies promises to essentially leapfrog the existing transportation infrastructure in undeveloped countries, much as cell phones have enabled developing nations to leapfrog landline telephone technology. This is one area where the IoT will make a world of difference.

 Note

The Matternet started as a project from a group of students at Singularity University, a "benefit corporation that provides educational programs, innovative partnerships and a startup accelerator to help individuals, businesses, institutions, investors, NGOs and governments understand cutting-edge technologies, and how to utilize these technologies to positively impact billions of people." Learn more about SU at www.singularityu.org; learn more about the Matternet project at www.matternet.us.

The Agricultural Internet of Things

While we're talking about rural areas, let's examine the IoT's impact on farming, both in the United States and around the world. As with all agricultural endeavors, the goal is to increase yields (by minimizing waste and improving productivity) while lowering costs and minimizing the effect on the environment. There are a number of ways that the IoT can help forward this goal, both for small farmers and large corporate agricultural operations.

Remember the one thing that the IoT is especially good at—monitoring things via its network of sensors. This has huge potential for application in agriculture, where the IoT can be used to monitor various aspects of the farming operation. Savings and efficiencies result, then, from analyzing the collective data, learning from it, and then acting on it.

Smart Irrigation

For example, up to 60 percent of the water used for irrigation is wasted. Smarter water management can result in less water used to obtain the same results. That's exactly what's promised by WaterBee, a smart irrigation and water management system that uses a network of wireless sensors in fields to collect data on soil content, water absorption, and other environmental factors.

WaterBee's system analyzes the collected data to enable farmers to selectively water different plots of land, as shown in Figure 14.2, based on need. (It can also be managed using a simple smartphone app.) It's estimated that this type of system can reduce water usage by up to 40 percent, which not only saves costs but helps alleviate water shortage problems.

 Note

WaterBee's system isn't limited to agricultural uses. It can also be employed for various commercial installations, including golf courses and vineyards. Learn more at waterbee.iris.cat.

Pest Control

Pest control is another big issue for farmers; in 2010, insect damage cost U.S. farmers $20 billion in damaged crops (and another $4.5 billion for insecticide). The challenge is to minimize the damage caused by insects and rodents without overly poisoning the land or crops with overly effective pesticides.

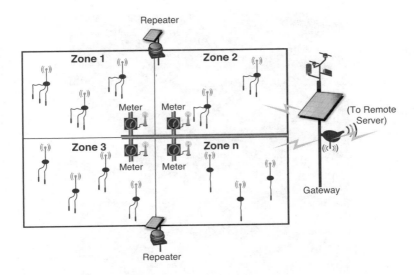

Figure 14.2   *How WaterBee's smart irrigation and water management system works, by monitoring conditions in separate plots of land.*

Enter Z-Trap, from Spensa Technologies, an electronic insect trap that enables farmers to remotely monitor their insect populations. As shown in Figure 14.3, Z-Trap uses pheromones to trap insects, then compiles data on the number of different types of insects in the trap. The data is wirelessly transmitted to the farmer's computer or smartphone, where the farmer can see a map of the type of insects that have been detected. With this data in hand, the farmer can then determine the best way to control those particular insects without "carpet-bombing" the entire area with a single pesticide.

 Note

Spensa Technologies is based in West Lafayette, Indiana, as part of the Purdue University Research Park. Learn more at www.spensatech.com.

Smart Tractors

Tractors of various sizes and shapes are a big part of the agricultural environment. Newer smart tractors include sensors that monitor data about crops and land, and adjust operations accordingly. A smart hay baler can sense the moisture content of the hay and then send signals to the tractor to move faster or slower through the field. (Moisture content affects bale density.) Or a tractor can use GPS technology to steer around waterways and other obstacles, and avoid duplicate seeding.

Figure 14.3   *Spensa Technologies' Z-Trap smart insect trap.*

As an example, John Deere's FarmSight system, shown in Figure 14.4, uses wireless technology to monitor individual tractors or fleets using computers, tablets, or smartphones over the Internet. The system provides data about the tractors themselves and the fields they travel.

Self-Driving Tractors

In the future, the concept of self-driving tractors (and other farm equipment) will no doubt take hold. This makes a lot of sense, as automating this particular operation cuts the costs and inefficiencies of human operation—and should improve operational effectiveness.

Witness the Spirit, shown in Figure 14.5, from the Autonomous Tractor Company. The Sprit is a self-driving tractor that navigates via a series of ground-based transponders installed around the perimeter of a plot of land. The tractor is also equipped with radar to detect other obstacles, from tree stumps to stray rabbits. A farmer "trains" the Spirit by leading around the perimeter so it knows where to go; if it later strays from the chosen path, it automatically shuts down.

Figure 14.4 *The cab of a John Deere S-series tractor, enabled with FarmSight technology. Learn more at www.deere.com.*

Figure 14.5 *The Spirit self-driving tractor from Autonomous Tractor Company. Learn more at www.autonomoustractor.com.*

The Environmental Internet of Things

Some benefits of the IoT are not limited to specific countries or locations. There is much the IoT can do to improve the quality of the global environment—even battling, to some small degree, the effects of climate change.

Much of what the IoT can do is more effectively monitor various environmental factors. (The IoT is essentially a network of sensors, remember.) For example:

- **CO2 sensors** monitor automobile emissions, pollution from factories, even toxic gases generated on farms. (Oh, those stinky pigs!)

- **Water sensors** monitor water quality in oceans, rivers, and lakes, and determine whether water is suitable for fish and plant life.

- **Radiation sensors** monitor radiation levels in nuclear power plants and in surrounding areas, and generate leakage reports and warnings when necessary.

- **Sensors in wooded areas** monitor the presence of combustion gases and humidity levels to better detect conditions that might lead to forest fires.

- **Electromagnetic sensors** monitor electromagnetic levels from cell towers, power lines, Wi-Fi routers, and other electronic equipment.

For example, the Smart Water wireless sensor platform from Libelium uses multiple sensors to measure a dozen or so water quality parameters. Sensors, like the one in Figure 14.6, monitor pH, dissolved oxygen (DO), oxidation-reduction potential (ORP), salinity, temperature, and dissolved ions. The various sensor nodes connect via the cloud to computers and smartphones to provide real-time water control.

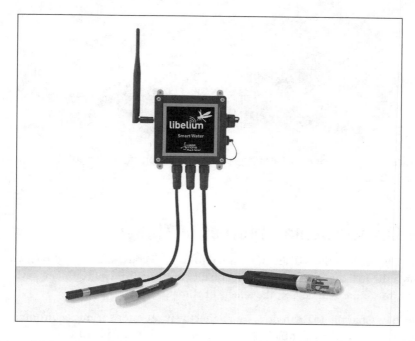

Figure 14.6 *Libelium's Smart Water sensor node. Learn more at www.libelium.com.*

If you want to monitor air quality in your immediate area, check out the Air Quality Egg, part of a community-led open source sensing network. Each Egg (available for $185 from shop.wickeddevice.com) collects nitrogen dioxide (NO2), carbon monoxide (CO), temperature, and humidity data in real time and then transmits that data over the Internet to a centralized website, shown in Figure 14.7. This website aggregates and displays data from every Egg in operation. The aggregated data can then be used to drive environmental policies.

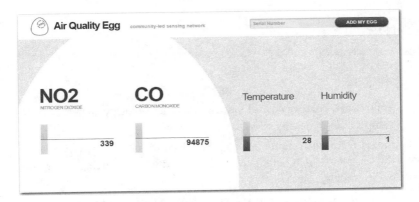

Figure 14.7   *Displaying data on the Air Quality Egg website (www.airqualityegg.com).*

Battling Climate Change

While we're on the topic of the environment, let's address the issue of climate change. In the past 50 years, the average temperature has increased at the fastest rate in recorded history. Arctic ice is melting, ocean levels are rising, weather patterns are shifting in none-too-subtle ways. Experts believe that the warming climate could have a devastating effect on our global food supply, which could be cut by 18 percent by the year 2050.

Climate change is real, and most scientists agree that it's largely caused by human-caused pollution. It's a big problem, it's our problem, and we need to solve it.

Enter the Internet of Things.

The primary way that the IoT can help battle climate change is by reducing global energy consumption, which in turn will reduce carbon emissions. We've already discussed various energy-saving implications of the IoT, but imagine all of these relatively local actions taking place on a global scale:

- **Smart thermostats** for more efficient heating and cooling of homes and businesses.
- **Smart lighting systems** for homes, businesses, public areas, buildings, and roadways that only turn on the lights when necessary.
- **Smart appliances and electronics** designed to run when overall energy usage (and costs) are lower.
- **Smart grids** that more efficiently route electricity to customers.

 Note

A recent study by Pacific Northwest Laboratory estimates that smart grid systems could reduce related carbon emissions by as much as 12 percent.

- **Smart cars, roadways, and parking lots** that reduce vehicle fuel consumption by reducing overall driving time.
- **Smart farming systems**, such as the irrigation systems discussed in this chapter, which help increase crop yield while reducing water, fuel, and pesticide usage.
- **Smart monitoring systems**, such as the Air Quality Egg, that help individuals and businesses better monitor air and water quality.

When it comes to using the IoT to fight climate change, it's a series of small changes that add up quickly. Consider, for example, the issue of traffic management. If you live in an urban area today, you're used to sitting in traffic during the morning and evening commutes; it's just part of the job. But all that idling in traffic jams wastes a lot of fuel—1.9 billion gallons a year in the United States alone. That translates to 186 million tons of unnecessary carbon dioxide (CO_2) emissions.

These carbon emissions could be dramatically reduced via the adoption of smart traffic management technology. If the IoT can help you avoid traffic jams and keep you moving toward your destination, that's less fuel wasted (which means more money in your pocket, of course) and less pollution emitted. Everybody benefits.

Impediments to the Global Internet of Things

If you think it's challenging to connect together a bunch of smart devices in your house to create a smart home, consider the challenges in trying to do just the same thing on a global scale. It's an increase in difficulty of several magnitudes.

When we're dealing with global IoT, however, the challenges are not all technological. Governmental bureaucracy and politics raise their not so lovely heads as well.

Technological Challenges

For the IoT to reach its full potential, there must be broad adoption across all market segments and geographic locations. You can't have smart homes, smart cities, and a smart world without all the necessary pieces and parts available and working in perfect harmony. If the parts don't fit together, the whole thing just won't function.

Imagine the traditional Internet being limited to the United States only, and you realize how important it is for smart devices to be adopted in all developed countries—and for all those countries to be connected to the same smart network. Devices in the United States must be able to communicate with similar devices in Canada, Mexico, France, India, Russia, and China. Exclude any country or region from the list and the usefulness of the entire IoT becomes compromised.

For the IoT to insinuate its way into civilized societies everywhere, there need to be standards that facilitate its adoption. Organizations in different regions and countries must come together to create the standards necessary for the technologies behind the IoT. Without standards, it's possible that we'll end up with different and likely incompatible IoT networks in different areas of the world.

If and when the necessary standards are set, there are further technological issues to address. One of the most significant is network bandwidth. When we're talking about adding tens of billions of smart devices to the current Internet, you have to wonder if the network infrastructure will be able to handle the increased load. The entire Internet backbone will need to be enlarged and reinforced, or new networking technologies developed to run alongside or outside the current Internet.

Of course, all the technology behind the IoT costs money. (That's why big companies like Cisco and Intel are salivating all over themselves to take the lead on all these smart things; they smell the money.) The problem governments worldwide face is coming up with that money. It's one thing for big corporations to devote part of their huge information technology (IT) budgets for IoT implementation; it's quite another for cash-strapped governments to do the same. Somebody has to pay for the IoT infrastructure, and it will probably be taxpayers like you and me. That might be okay in relatively affluent Western nations, but developing countries will have a harder time coming up with the funds to jump on the IoT bandwagon.

Security Challenges

As we'll discuss in Chapter 15, "Smart Problems: Big Brother Is Watching You," the more devices you connect to the global network, the more points you have for hackers to enter and corrupt the system. Billions of new devices mean billions of new opportunities for the bad guys to exploit. That's a challenge.

In addition, all the data collected by all these new IoT devices puts a lot of personal information out there just waiting to be stolen or legally abused by those in power. The IoT devices, when up and running, will know quite a lot about you and what you do; who will have access to that information, and how will they use it? Even if you think that's not such a big problem in the United States (and that is extremely debatable), there's still the issue of what less altruistic governments in China, Russia, or North Korea might do with the same information. It's a scary situation.

Then there's the issue that the more we rely on the IoT, the more vulnerable we'll be if some or all of it is attacked or goes offline. If we rely on the smart grid to manage all our electricity use, what do we do when North Korean hackers (or just that weird kid down the street) hacks into and takes control of the smart grid systems? How about if a malcontent hacks into your smart car's autonomous guidance system? Or terrorists take control of Domino's fleet of pizza delivery drones?

For these reasons and more, we need appropriate safeguards to ensure that our smart devices don't get hacked into and the IoT doesn't fall under malicious control. Otherwise, the consequences are frightening to consider.

Bureaucratic and Political Challenges

You think it's hard to get a purchase decision through the various committees and levels of management in a typical corporation? Then consider the same scenario on the level of city, state, or national government. Even the smallest decisions are rife with bureaucratic impediments and political ramifications.

It would be naïve to think that the adoption of the IoT on a national or global level would be immune from these challenges. In a world where it takes years of studies, commissions, meetings, and political maneuvering to get a stoplight installed at a busy intersection, imagine the challenges in getting our elected (or, in some areas of the world, self-appointed) officials to agree on something as conceptually and technologically sophisticated as the IoT. These are the folks who think the traditional Internet is a series of tubes, remember; there's no way some of them will ever understand how the more complex IoT works.

That said, many of our officials do understand both the potential benefits and the challenges represented by the IoT. We have to rely on these technologically astute lawmakers to take the lead and push forward the IoT agenda.

Some of this can be done on a local level, in the course of day-to-day operations. Public building managers should employ smart building, energy, and lighting technologies. State and city lawmakers should invest in intelligent transportation, and waste and water management systems, and ensure that public utility commissions deploy smart meters and similar technologies. Federal policymakers should ensure that funding for highways and bridges includes money for smart sensor technology, and that regulatory agencies are prepared to review and adopt innovative IoT technologies. It's a matter of leading by example to influence the eventual public debate.

THE SMART WORLD AND YOU

How will this impending global Internet of Things effect you? You're just a simple guy living in his house in Peoria (or a gal in her apartment in Portland)—what do you care about smart irrigation systems in India?

First, we all live on the same planet, and we ought to care about what happens outside of our immediate field of vision. If we can use technology to make life better for the residents of Bangalore or Budapest, why shouldn't we?

Second, smart technology isn't limited by national borders, and neither should be our thinking. The car you drive was manufactured in Japan (or Korea or Germany) and contains parts from subsuppliers located in various countries around the world. If its smart devices need to phone home to the automotive mothership to effectuate various functions, it's important that the global systems exist to facilitate that communication. It's not just your Internet of Things, it's everyone's—all around the world.

Third, and back to the "we're all in this together" argument, what happens in India, China, and Australia affects the global environment. Climate change is the best example of these interdependent relationships—one single country cannot battle climate change by itself; all countries have to work together to make an impact. And, as much as the IoT can help address this and other issues, we need to work on implementing it not just in America, but around the world.

Fortunately, other countries are recognizing the potential promised by the IoT and working on their own implementations. Take China, for example. Already, China accounts for more than a quarter of the world's machine-to-machine (M2M) connections. That's more than 50 million smart devices in that country, due in no small part to the strong support of the Chinese government, which plans to invest more than $600 billion in IoT technology by the year 2020. You see, China knows what's what, so it is investing in the technologies required to make its homes and cities smarter—and provide a better quality of life for its citizens.

So when will this global Internet of Things be ready for prime time? Obviously, adoption rates are going to differ from country to country, but if you want a good guess, most experts expect that the IoT will begin having a major impact somewhere around 2025.

That's about a decade from now, which isn't that terribly far into the future. This is something you will see in your lifetime, and that your children and grandchildren will grow up with. It's going to be big, it's going to be important, and it's going to be everywhere. Get ready for it!

15

Smart Problems: Big Brother Is Watching You

After 14 chapters of rose-colored predictions for the Internet of Things-enabled future, it's time to discuss the flip side of all this smart technology. You know, where all the good things about the IoT get used against you.

Most of what's potentially dangerous about the IoT concerns the data collected by the IoT's smart devices. You see, the companies that sell you all those nifty smart devices will eventually know pretty much all there is to know about you. What you do in your home, in your car, at work, in the store, and within your community will be increasingly monitored and analyzed in ever more intrusive ways by corporations who want to sell you more stuff. It's also possible—likely, actually—that this information will be accessed by the government. And who knows what they will do with it.

The upshot is that, thanks to the IoT, more and more of what used to happen behind closed doors will be open to scrutiny and exploitation by parties that you would never willingly invite into your home. If that makes you feel a little uneasy, good. It should.

Privacy Issues

Probably the biggest issue with the IoT concerns privacy. All those sensors and smart devices will be collecting a lot more information about you—what you do, when, where, and how. Who has access to that data, and what can they do with it?

Let's face it, we're living in a golden age of surveillance. It's worse in some countries than others, but we're constantly being monitored, both in public and private. In cities big and small, you're never more than a few steps from a surveillance camera. Global positioning system (GPS) chips in your smartphone track your every movement. The government assumes it has the right to read all your emails and text messages, as well as track every website you visit, whether you give permission or not. It's almost impossible to go "off the grid" and avoid the watchful eyes of Big Brother; if someone or something wants to track you, they can and will.

Of course, some of this is a tradeoff. You want that online retailer's website to remember your preferences and what you ordered last, so you don't have to reenter that information every time you visit; the tradeoff is that the site installs a cookie on your computer that tracks what you do on that site, as well as where you came from and where you go when you're done. You want security from muggers and rapists and terrorists, which security cameras provide; the tradeoff is that you get monitored along with the bad guys. You get some benefit from a given technology—you also have to pay for it.

What Do They Really Know About You?

One of the main reasons we're looking forward to the IoT—and embracing those smart devices currently on the market—is the ability to gain useful data about ourselves and our surroundings. We want to monitor our home energy usage, how many steps we take in a day, or how many eggs we have left in the fridge.

This information, useful to us, can also be valuable to many other people and companies. Of primary interest are the companies who make these smart devices, who can profit by selling the data they collect to other parties. You know that the business plans of many of these smart device manufacturers includes a revenue stream from the sale of collected data.

Now, many of these companies say they'll only sell aggregated anonymized data— that is, the numbers from many customers all added up, not the specific details about you or other individuals. Even if that is true (and not all companies are saying they'll do this), there's still the possibility, if not the likelihood, that individual details can and will be extracted from the whole.

All this data can be analyzed to reveal a variety of intimate details about what you do during the day, where you go, your interaction with others, your medical needs, your personal habits. You may be comfortable with all this information being available to strangers; many others may not.

How bad can it be—what kind of data are these smart devices going to collect, anyway? Let's look at a few examples.

Consider a smart thermostat like the Nest Learning Thermostat. This little hockey puck collects a huge amount of data that could be of interest to various parties. Yes, it learns your heating and cooling preferences, but that information isn't terribly interesting. What's more interesting is the data that other devices share with the Nest. When the thermostat is connected to your garage door opener or car, for example, it knows when you've left the house, so it can enter into "Away" mode. Who might want to know when you've just locked the door and driven out of the driveway? Burglars, of course. Or maybe bill collectors who want to grab that expensive lawn mower on which you're three months behind on payments. Or even an ex- or soon-to-be ex-spouse who wants to build a case about your philandering ways.

What about your sex life? Like to keep it private? That won't be quite so easy when the IoT takes hold. Take, for example, the upcoming Sexfit penis ring from Bondara, which is kind of like a pedometer for your private part. (We'd show a picture of it here, but we're trying to be decent for the kiddos.) In addition to its "LED performance indicator lights" and "personal trainer vibration mode," it also transmits data via Bluetooth to your smartphone, where the corresponding app displays your thrusts per minute and calories burned. You want that data to be hacked?

 Note

In a related matter, some Fitbit fitness band users have found statistics about their sexual activity posted online. Not cool.

Back to what's happening in the clean world, there's a lot of interesting data that can be gathered from your smart car, too. Like where you're at every minute of every day. Do you ever visit a bar or establishment that your spouse might not approve of, or your boss might have adverse thoughts about, or your pastor might consider sinful? Well, that data is going to be collected and could be used against you.

Some uses of this collected data aren't quite as nefarious but could prove equally annoying. Your smart TV knows every program you watch; there are lots of companies who'd like to have that information—the better to feed you advertisements based on your viewing habits. Shopping data is equally interesting to these parties, as is data that can infer other leisure activities. Or maybe a home goods company buys your personal data from your smart thermostat company and uses it to pitch you heavier blankets in the wintertime. It's not criminal, but it is insidious.

 Note

When data is collected and stored, it also becomes vulnerable to theft. Even if the company that collects the data doesn't do anything untoward with it, a hacker breaking into the company's database will find data that he can personally use or sell to another criminal party. Just having the data out there is a risk.

Your Government Is Spying On You

Then there's the government, those elected and appointed officials in charge of managing this great city, state, or nation of ours. There's a whole apparatus in place that's charged with keeping a watchful eye on us citizens, law-abiding or otherwise. We're talking the FBI, the CIA, the NSA, the whole Department of Homeland Security. It took an Edward Snowden to reveal the extent to which the U.S. government spies on its citizens. Although some restrictions are theoretically in place, the spooks in Washington pretty much have carte blanche to listen in on every telephone conversation, read every text message, and pour through the entire contents of your email inbox. There's nothing you can do on the Internet that they can't monitor.

That's today. Imagine the field day they will have with the reams of data collected by tomorrow's smart devices. Billions of devices connected to the IoT means billions of new things the government can track about you and your neighbors. They might say they're doing it for our own good, to protect us from nebulous or nonexistent terrorist threats, but it's spying on us all the same. It's all very Big Brotherish; other than the year, it appears that Orwell had it dead to rights.

This isn't idle speculation, either; the government is practically salivating in expectation of the surveillance bonanza on the horizon. In 2012, former CIA director David Petraeus spoke at a summit for In-Q-Tel, the CIA's venture capital firm, about how the IoT is going to change the world of legalized spying. "'Transformational' is an overused word," Petraeus said, "but I do believe it properly applies to these technologies, particularly to their effect on clandestine tradecraft."

I like that phrase, "clandestine tradecraft." It's much nicer than saying "spying on citizens." But let's not get sidetracked by language. There's a serious issue here.

In the past, if the CIA or other agencies wanted to track your movements, they had to put a couple of cars full of agents on your trail and physically follow you as you drove around town. With the rise of smart technology, all they have to do is intercept the data from your smart car, smart home, and smartphone to know exactly where you are at any point in time, and probably what you're doing there, too. The IoT serves as the data collection service; the spooks just have to access that data. It makes their jobs a lot easier.

Petraeus also had this to say:

"Items of interest will be located, identified, monitored, and remotely controlled through technologies such as radio-frequency identification, sensor networks, tiny embedded servers, and energy harvesters—all connected to the next-generation Internet using abundant, low-cost, and high-power computing."

That's as good a description of how the IoT works as any I've seen. The CIA certainly is on top of the technology. And it knows that these smart devices will, again in Petraeus' words, "change our notions of secrecy."

Privacy Versus the IoT

How much privacy do you have a right to, or want, or need? One can argue that that's just the tradeoff, a line that's been long crossed—your privacy in return for all the wonders of the digital world. In today's electronic age—and tomorrow's world of the IoT—privacy simply isn't part of the equation. In order to benefit from the interconnectivity of the IoT, privacy is willingly abandoned.

In fact, many believe that the IoT will be an age of "radical openness," where every transaction, every operation, every person, and everything will be totally transparent. That transparency is necessary to establish the type of trust necessary to allow all your devices to connect to all the other devices out there; you have to trust that things will work and that they will work on your behalf.

Others, privacy advocates especially, believe that IoT companies should not collect this data, even anonymously. Privacy advocates want transparency, too—they want manufacturers of smart devices to be perfectly clear about what data they collect and for what purposes. Consumers should have the option of turning the data flow if they want or deciding who gets access to that data.

The problem with turning off the data flow into the IoT is that it turns smart devices back into dumb ones. It's exactly that interconnected flow of data that makes the IoT so intelligent; if the data isn't flowing, all you have is a collection of isolated sensors.

There are no easy answers to this one, other than to reiterate that there are trade-offs involved. If you want total privacy, you may not be able to benefit from the IoT—and if you want to benefit from the IoT, you may have to sacrifice some of your privacy.

 Note

If you're at all concerned about your online privacy, look into the work being done on your behalf by The Electronic Privacy Information Center (www.epic.org) and the Privacy Rights Clearing House (www.privacyrights.org).

Security Issues

Related to the IoT's privacy issues are issues that have to do with security. These fall into two broad camps—how secure is your own data on the IoT, and how secure is the IoT as a whole?

Data Security

Let's assume, for the time being, that all the personal data collected from your smart devices is not being used at all by the companies that manufacture those devices, or by any governmental agency that thinks you're a "person of interest" in any way, shape, or form. That doesn't mean your data is safe; hackers could still break into the network and steal your personal information.

Of course, hackers can and do break into existing systems to steal personal information. Witness the 2014 attacks on Home Depot, Target, and a larger number of retailers than you'd like to remember. (Or, for that matter, the 80 million personal records stolen from Anthem, the country's second-largest health insurer, in early 2015.) You don't need the IoT to present an attractive target to criminals who want to abscond with credit card numbers, social security numbers, and other valuable data.

But the IoT does represent a larger and even more attractive target. First of all, all those billions of upcoming smart devices are all possible access points for malicious intrusions. If a hacker can't break into your utility's customer database, maybe he can hack into your own personal smart meter and access your data (or the main customer database) that way. Every smart device, every sensor, every input or output point can be a potential gateway into the system for a skilled and persistent hacker. And, given that many of these individual smart devices don't necessary have the same level of security protection as the big systems do (or even what you have on your current personal computer [PC]), these are easy targets. Very easy targets.

In addition, all the data collected by these smart devices gives the bad guys that many more reasons to steal it. Today, they want your credit card numbers; tomorrow, they may find a lucrative aftermarket for information about your television viewing habits, automobile usage, or even physical activities. Every bit of data is valuable to someone, which might make it worth stealing.

System Security

Some attackers have even more malice in mind than just stealing a bunch of digital data. Think of the problems that could ensue if hackers gained control of the smart devices in your home, car, or city.

We're talking cyberterrorists who break into the IoT (or some subnetwork therein) with the sole purpose of gaining control of important systems and operations. When the smart devices in your home are no longer under your control, mayhem can result.

Admittedly, some of these scenarios might sound comical. A cyberterrorist gaining access to your home's smart lighting system could turn your lights on and off at random. Someone hacking into your smart TV could feed you unwanted commercials or propaganda broadcasts. A bad guy hacking into a smart toilet might make it flush repeatedly and force the lid to keep going up and down.

Okay, not too scary. But there are more ominous scenarios. How about a hacker breaking into your smart lighting and security system to kill all your lights and alarms, and unlock all your smart doors, in preparation for a robbery or home invasion? Or a cyber voyeur hacking into your smart security system to spy on you via your smart cameras? Or someone with even more malicious intent turning a company's heating/cooling system against them by cutting off air flow or inducing dangerous gases into the system?

 Note

There are already numerous instances of hackers taking over Web-based baby monitors and webcams to spy on unsuspecting homeowners.

Consider the havoc that could result from having your smart car hacked. A bad guy could shut down your smart car remotely, or force it to turn into oncoming traffic, causing inconvenience, injury, or death. All it takes is one smart techie with a grudge, and a less-than-secure smart vehicle could become a death trap. .

It gets worse when you consider the larger city-, state-, and nationwide smart systems under development. If you think the North Koreans' cyberattack on Sony was a bit of work, what if they decided to attack your friendly neighborhood power plant instead? Or tried to take control of your local water company? Or the nation's arsenal of nuclear weapons?

The more things we connect via the IoT, the more things that malicious individuals or organizations can try to damage or control. It may sound nice to have virtually every device in the world connected via the IoT, but it makes for a very scary situation security-wise.

What's the solution? More and better security, as always. Some of this is on you, the consumer, but most is on those companies collecting, transmitting, and managing the data generated by IoT devices. Every point in the network needs to be secure, which is a daunting task. The IoT network, as large as it likely will become, will only be as strong as its weakest link—that is, the least protected smart device.

Big Data Issues

Speaking of data, what if we end up collecting *too much data* to be practically useful? If there's too much data filling our collective unconscious, do we experience paralysis by analysis?

What we're talking about some people call *big data*, extremely large data sets that can be analyzed to reveal patterns, trends, and associations otherwise not easily visible. That sounds great, and it's part of the appeal of the IoT—making things more intelligent by tying together more data from more devices.

It's possible, however, to collect so much data that we don't have the time, the facilities, or the computational power to analyze it all. We end up collecting data for data's sake and never doing anything with it. It just sits somewhere, out in that omni-present cloud, taking up storage space. Forever.

If this were to happen, then it's a huge waste of a promising technology. Yes, future smart devices will still likely talk to each other on some low level, but the potential represented by this universal network of data collection points will never be realized.

What this points to is the need for data analysis in the form of smart systems and even smarter human beings. If the IoT is to truly impact our future lives, we need to train an army of data analysts. We need people who can intelligently, accurately, and creatively analyze the data collected from all the different sources that will make up the IoT. The data itself isn't intelligent; it's the analysis that provides intelligent interpretation.

We also need people to devise the intelligent systems that also analyze and act on the collected data. Programmers who can devise the right algorithms to make sense, in near real-time, of all the reams of data from various sources. While we can't rely on automated systems and algorithms to fully analyze all the IoT data, we can't rely on human beings to work through it all either. We need a combination of human experts and expert systems to make it work, or all we're doing is collecting a bunch of garbage.

Autonomy and Control Issues

The influx of smart devices and systems makes for a bit of a moral dilemma as well. How autonomous do we want our autonomous devices to be?

Now, some of this is a matter of culture and human perception. Human beings, in general, like to be in control of things. The more control we cede to our devices and machines, the less we feel as if we're still in control. That can lead to feelings of vulnerability and even inferiority. When we feel as if we're not in control, we're less sure of ourselves and have lower self-esteem. Control is important.

So what happens when we give more and more control of our lives to things and systems that we ourselves do not control? Do we really want to do this—can we really do this? Or is there some line that we as a society simply won't cross, some point at which we want to keep control, no matter how illogical that might be?

Consider the self-driving car. The concept sounds dandy for a lot of people, but how comfortable will you really be when the day comes to sit back and turn the driving over to a machine? Personally, I like to drive—I'm a much better driver than I am a passenger. I'm not sure I want to turn that function over to an automated system in my car. I want to be in control, and I'm not about to figuratively pass the steering wheel over to a machine. It's a control thing.

Obviously, there are many things in life that we are comfortable turning over the control to. Some functions are just too mundane to bother with; we willingly let someone or something else handle them for us. But can we do that with virtually *everything*? Will we need to maintain some control over those systems that are, in essence, controlling us? Or will we become like the blob-like future humans in the film *Wall-E*, lounging on hover chairs while their every action is controlled by their intelligent spaceship host?

Smart Machine Issues

Speaking of autonomous devices, what do we do when our smart devices and systems get a little *too* smart and decide they don't need us puny humans around?

Now we're into the realm of science fiction, or so we hope. Turn no further than the *Terminator* movies, in which the Skynet network of smart machines gained self-awareness and turned against its human creators. Many bad things then happened, many of which involved Arnold Schwarzenegger.

This notion of machines or systems becoming self-aware and gaining true intelligence manifests itself in the concept of the *technological singularity*. This is the theoretical point at which artificial intelligence will exceed human intellectual capacity—and control. Futurist Ray Kurzweil projects that given the current rate of technological advancement, the singularity will occur somewhere around the year 2045. That's something to look forward to.

What happens if and when the singularity is reached? Perhaps our smart houses and smart cars will become our intellectual equals, much like Tony Stark's J.A.R.V.S intelligent computer in the *Iron Man* movies and comic books. Or maybe they'll become our rivals, as in the *Terminator* franchise. Or perhaps they'll be benevolent helpers—until their needs and programming take precedence over ours, as happened with the HAL 9000 computer in *2001: A Space Odyssey*. I'll leave the debate to others with a firmer grounding in either science or science fiction— although I'm sure Asimov's laws of robotics will enter into it.

 Note

The late, great science fiction writer Isaac Asimov postulated the Three Laws of Robotics in his 1942 short story, "Runaround" (and later collected in the book, *I, Robot*). They are, in order:

(1) A robot may not injure a human being or, through inaction, allow a human being to come to harm. (2) A robot must obey orders given it by human beings except where such orders would conflict with the First Law. (3) A robot must protect its own existence as long as such protection does not conflict with the First or Second Law. And thus the human race is protected against rogue robots it might inadvertently create.

The point is, the more autonomy we cede to our devices, systems, and machines, the more this scenario becomes possible, if not probable. We're deliberately creating autonomous intelligent systems, so it should come as no surprise if and when they gain true intelligence and autonomy. Whether these future intelligent machines become our friends, helpers, or enemies depends a lot on how we program them. Let's hope there's some kindness involved.

SMART PROBLEMS AND YOU

How excited should you be about the Internet of Things? That depends on how much you accept the prevailing premises and how much you worry about the potential problems.

On the plus side, there's a lot to look forward to. Smarter devices and smarter systems promise to automate a lot of the boring and mundane processes and decisions in our lives. By connecting all these devices together, smarter, more efficient decisions and operations should result. The world will become a better place; you'll have more leisure time, fewer worries, and more money in your pocket. Everything is good.

On the downside, all that data collected by those smart devices could be used against you by unscrupulous advertisers, malicious hackers, and our friends in various government agencies. Personal privacy could become a thing of the past, and it's even possible that all these connected systems (personal and public) could be hijacked for malevolent purposes. Hell, it's even possible that our smart systems could rise up against humankind and destroy civilization as we know it. Possible, if unlikely.

The point is, every step forward holds both promise and concern. Every technological advance has the potential for both good and evil—or just the mundane. It's not just automatically going to be all pie-in-the-sky goodness and light.

Some problems inherent in the IoT are going to be out of your control. You have little to no control over the smart grid, for example, or the adoption of smart highways or smart waste collection. Somebody further up the food chain is making those decisions, and you'll have to live with them.

At your level, however, there are some decisions to make. You can choose whether or when to embrace smart devices in your home. You don't have to install smart lighting or smart security systems in your home; the old-fashioned ones work just fine, thank you. You don't have to connect your home to the smart grid or purchase a smart car if you don't want to. That's your choice.

That doesn't mean you can't be involved in the larger decision-making; that's what elections are for. Keep informed about the emerging technology and issues (you already are, by reading this book) and enter into a public debate. If you see the benefit, support development. If you fear the consequences, argue against adoption—or for more stringent controls.

You should also support your views with your pocketbook. The Internet of Things will succeed or fail based on what happens in the marketplace. If enough consumers (individuals or companies) buy the new devices, the IoT will become a real thing. If consumers stay away in droves, the whole thing will be deemed a failure and nobody will remember it ten years from now.

In other words, the triumph of the Internet of Things is not preordained. Technology marches on, yes, but not always in ways that are readily apparent at the time. Back in the early 1990s, nobody really predicted that the Internet would evolve into a medium for watching movies, buying goods and services, exchanging gossip, and viewing pornography. (Okay, maybe the porn was predictable.) There's no way that we can predict today what the IoT will evolve into ten or twenty years hence. It may become the greatest thing since sliced bread, or it may end up being nothing more than yesterday's leftovers.

Should you be concerned about the potential problems presented by the IoT? Of course. Should you immediately disconnect all your devices, smart or otherwise, and stock up supplies in your bunker to carry through the coming war with the machines? Probably not. Worse comes to worst, it's not going to be that bad.

In fact, it's likely that the IoT will live up to some if not a majority of its promises. Not without hiccups, of course, but nothing ever runs completely smoothly. Stay on top of what's happening, support the good stuff, guard against the bad stuff, and everything will likely turn out all right. After all, tomorrow is just your future yesterday—and yesterday wasn't all that bad, was it?

Sigfox ?
thethingsystem.com

Index